Other **WILD**Guides titles

 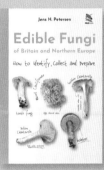  

press.princeton.edu/wildguides/

HOVERFLIES
of Britain and Ireland

Stuart Ball and Roger Morris

Main contributing photographers
Steven Falk and Paul Kitchener

WILDGuides

PRINCETON
UNIVERSITY
PRESS

Published by Princeton University Press,
41 William Street, Princeton, New Jersey 08540
99 Banbury Road, Oxford OX2 6JX
press.princeton.edu

First published 2013
Second Edition 2015
Third Edition 2024

Copyright © 2013, 2015, 2024 by Stuart Ball and Roger Morris

Copyright in the photographs remains with the individual photographers.

All rights reserved. No part of this publication may be reproduced, stored in a retrieval system, or transmitted, in any form or by any means, electronic, mechanical, photocopying, recording, or otherwise, without the prior permission of the publishers.

Requests for permission to reproduce material from this work should be sent to permissions@press.princeton.edu

British Library Cataloging-in-Publication Data is available

Library of Congress Control Number 2023930513
ISBN 978-0-691-24678-9
ISBN (ebook) 978-0-691-24679-6

Editorial: Andy Swash and Megan Mendonça
Design: Rob Still
Cover Design: Rob Still
Production: Ruthie Rosenstock
Publicity: Caitlyn Robson-Iszatt and Matthew Taylor

Cover image: Male *Sericomyia silentis* by Martin Jones
Title page image: Female *Rhingia campestris* by Ron James

Printed in Italy

10 9 8 7 6 5 4 3 2 1

Contents

Foreword 5
Preface 6
Introduction 8
Is it a hoverfly? 10

Hoverfly biology 12
The life-cycle of a hoverfly 12
Adults 13
Eggs 15
Larvae 16
Pupae 26
Migration 26
Polymorphism and other colour variations 28
Mimicry 30

Finding hoverflies 32
The seasonal calendar 33
Where to look for hoverflies 38

Glossary 44

Identifying hoverflies 47
Naming the parts 48
Guide to the tribes 53
Simplified guide to British hoverfly tribes 61
Identifying wasp and bee mimics 62
A guide to the most frequently photographed hoverflies 66

Introduction to the species accounts 70

THE SPECIES ACCOUNTS 72

List of British and Irish hoverflies 302
Photographing hoverflies 311
Collecting hoverflies 317
Legislation and conservation 322
Recording hoverflies 323
Putting data to good use 324
Research opportunities 333
Gardening for hoverflies 334
Further reading and useful addresses 337
Acknowledgements and photographic credits 339
Index of scientific names 342

♀ *Volucella inanis*

Foreword

Hoverflies have long been regarded as nice insects, charismatic even. And they get a good press as our friends, not only because they are colourful and tamely sit on flowers, but because many are useful to gardeners in controlling the numbers of aphids attacking plants.

Yet despite this popularity, many people would not be sure whether they are looking at a hoverfly or not. Many species are excellent mimics of bees and wasps, and such is the variety of appearance that many other species would not be recognised as hoverflies at all.

One of the big incentives to looking at hoverflies has been the rapid rise in popularity of close-up digital photography. Many people are discovering a new wonderful world of small creatures which are amazing when seen in close-up detail on a screen. That is step one, but the enquiring mind then asks, "but what is this marvellous fly and what does it do for a living?"

This is where the Hoverflies **WILD**Guide becomes invaluable. A set of high-quality photographs of live insects is designed to facilitate identification, accompanied by information on life history and distribution. The photographs also illustrate the angles of view that best show identification features.

Many other books include a small selection of example hoverflies, in some cases wrongly identified or with erroneous drawings. This Princeton **WILD**Guides publication is unique in that it covers all the genera and a high proportion of the British and Irish species, with accurately named photos, and is authored by the co-organisers of the Hoverfly Recording Scheme.

There is not space to cover every species recorded in Britain and Ireland, and, indeed, on average about one new species is added to the list each year. Most of the omitted species are either unlikely to be encountered or cannot be accurately identified on the basis of a photograph. Full coverage is provided in the mini-monograph *British Hoverflies* published by the British Entomological and Natural History Society.

Hoverflies are one of the most diverse and fascinating groups of insects in their range of life styles. This makes them an especially important group of animals for monitoring the health of the countryside, and indeed town, and for monitoring biodiversity responses to climate change and various other environmental issues. But the main purposes of this **WILD**Guide are to help you to share in the pleasure of looking at hoverflies and to provide encouragement to contribute to knowledge of our fauna via the Hoverfly Recording Scheme.

Alan Stubbs

Preface

When we first set out our ideas for an entry-level guide to hoverflies more than ten years ago, we had no model upon which to base it. The concept was entirely new, and we could only guess how it might be used and what its impact would be. Also, whilst we had several decades of experience working with the British monograph (Stubbs & Falk: *British Hoverflies: An Illustrated Identification Guide*), and had spent many years running training courses, we were also novices when it came to photographic identification.

Today, after two editions of this **WILD***Guides* field guide we have a lot more experience upon which to draw. We have learned a great deal from working with photographic recorders through various social media, and know a lot more about what can, and cannot, be identified from photographs. In particuliar, we have a much more detailed picture of the mistakes people make. Furthermore, we have benefitted greatly from the contributions of specialists working in continental Europe who have drawn attention to additional useful features and the need for greater rigour in applying relevant characters to photographic identification. Hoverfly identification has advanced considerably over those ten or more years.

Our original idea was to develop a book that would act both as an introductory guide and as a companion to Stubbs & Falk. It is noticeable that in many ways this **WILD***Guides* field guide has replaced Stubbs & Falk as the 'go-to' guide book: numerous high-impact papers

♀ *Episyrphus balteatus*

quote our book, which is a great compliment, but we worry that academia has forgotten that high-end research also demands high-end taxonomy and the use of detailed keys to confirm the identification of many species. Thus, in refining the book for a third edition, we have continued to avoid detailed coverage of all taxa. Making critical identifications of difficult taxa remains the preserve of the microscope and pinned specimen. Hoverflies are far more accessible today, but they are by no means easy to identify. Although some can be recognised from photographs, in many cases identification is only possible from the highest quality photographs taken from several angles. We remain very concerned that traditional taxonomy is on the wane, even though there is a growing need for taxonomically competent biologists to undertake the identifications needed to inform the growing programme of DNA analysis.

We would like to think that this book has at least contributed to the growing popular interest in hoverflies. When we embarked on the project, the Hoverfly Recording Scheme compiled around 25,000 records each year. Now, the number of records has risen to more than 100,000 annually! The book is part of the reason, but social media, especially Facebook, have totally changed the paradigm. There is now a vibrant on-line society of hoverfly enthusiasts who are making a real difference. We know a lot more about hoverfly ecology because of their work; they have also added new species to the British list and there is rapidly growing interest in larval ecology.

This new edition tries to capture some of the lessons learned since the second edition was produced in 2015. It includes 12 additional species and extended coverage for two others that were mentioned previously but were not the subject of a full species account. The maps now include Ireland, although these should be considered provisional in view of the relative sparsity of records from Ireland when compared with the number from Britain. Species accounts have been revised where necessary to highlight new identification features and to include comments on distribution where appropriate. By necessity, with space at a premium, these accounts focus on the key identification features and on providing caveats to help avoid mistakes. In addition, many new and replacement photographs have been included to better illustrate what observers are likely to see and photograph, or to enhance the overall quality of photographic reproduction.

We are hugely grateful to the many hoverfly enthusiasts who have contributed to the growing knowledge of the British hoverfly fauna, and especially to those people who have contributed to the wealth of photographs in this new edition. Their contributions are recognised specifically in the acknowledgements section. We would also like to thank several European friends and colleagues for their help with photographic identification over the years who have provided much-needed advice on various occasions. We are very aware that ours is an island fauna, and that British and Irish enthusiasts cannot work in isolation from mainland Europe and beyond. There is a growing international community of hoverfly enthusiasts and we would like to think that in some small way the original WILD*Guides* field guide was one of the stimuli that has facilitated its development.

Stuart Ball & Roger Morris
February 2024

Introduction

On any fine day between April and October, whether you are in the countryside or in an urban or suburban park or garden, you are likely to come across brightly coloured, black-and-yellow flies hovering around flowers. Although they are trying to convince you they are wasps or bees, they are actually hoverflies. They are such constant flower visitors that in some other parts of the world they are called 'flower flies'.

At the time of writing, 285 species of hoverfly are listed for Britain and Ireland, with more being discovered at a rate of approximately one species per year. The bright colour patterns of some make them readily recognisable but others are not easy to identify in the field and require microscopic examination to be certain. This book focuses on identification of the commoner and more distinctive species. In total, 177 species are illustrated and described, covering mainly those that are easiest to recognise, even if they are rare or difficult to find. In order to show the full variety of hoverflies, all 69 genera occurring in Britain and Ireland are included. The sections on **Identifying hoverflies** on *page 47* and **Further reading** on *page 337* explain where to go next if you want to tackle some of the more challenging species.

In order to appreciate fully the relationship and similarities between species you need to understand how hoverflies are classified. The **Guide to the tribes** on *page 53* attempts to shed some light on what can be a confusing situation: becoming a competent observer of hoverflies depends upon a comprehensive understanding of the characters that are used to separate individual species.

This book uses a combination of field photographs and close-up digital images to show in detail the characters used to separate the species, and icons are used to provide an indication of the difficulty of identification. It includes up-to-date distribution maps and diagrams showing the flight-periods. These are derived from the Hoverfly Recording Scheme, the National Biodiversity Data Centre in County Waterford, Ireland, and CEDaR in Northern Ireland (see *page 323*). Information on how the maps were prepared is provided on *page 71*.

Digital photography has allowed many more people to take pictures of insects in the field in order to identify them back at home. There is a growing trend for this to be regarded as an alternative to the collection of specimens, but photography remains just one part of a bigger process of making an accurate identification; the species accounts indicate where photographs can be used reliably for this purpose. The arguments around the ethics of collecting specimens are covered on *page 317*.

Apart from being attractive and interesting, hoverflies also play important roles in the environment. Most gardeners know that hoverfly larvae are voracious predators of aphids and are, therefore, 'friends'. However, there are two species (the Greater Bulb Fly *Merodon equestris* and Lesser Bulb Fly *Eumerus funeralis*) whose larvae tunnel in daffodil and other bulbs and are therefore not so welcome in the garden. Another species, *Cheilosia caerulescens*, which was first recorded in Britain in 2006, will also be highly unwelcome to some gardeners as its larvae mine the leaves of houseleeks. It is, however, regularly reported by people who record the hoverflies in their garden.

Most hoverflies do not have common names. There are a few that do, however, including the two mentioned above, as well as the Drone Fly *Eristalis tenax* and the Marmalade Hoverfly *Episyrphus balteatus*. Common names catch on when they sum up some aspect of the species' appearance, behaviour or habitat in a way that is memorable. Contrived names

Male Greater Bulb Fly *Merodon equestris*   Male Drone Fly *Eristalis tenax*

seldom manage this and often end up being no more memorable than the scientific name they try to replace. Common names can also add to misidentification problems because the same name has been given to different species *e.g. Eristalis cryptarum* and *Sericomyia silentis* have both been named the 'Bog Hoverfly', one is vanishingly rare and the other is widespread, so thankfully the majority of misreporting can be rectified. For this reason, although the few established common names are included where appropriate, scientific names have been used in this book.

Hoverflies are sensitive indicators of the health of our environment because they are short-lived, fast-breeding and show rapid changes in both range and abundance in response to change: these responses can be detected and interpreted. Some of the changes that have been revealed by analysis of records collected by the Hoverfly Recording Scheme are discussed later in this book and in the species accounts.

Hoverflies are arguably the most attractive and accessible group of flies, so you might expect that there would be plenty of popular literature about them – but this is not so. The standard monograph is *British Hoverflies* by Alan Stubbs and Steven Falk, published in 1983 and fully revised in 2002. Two European guides are essential literature for the specialist: *Hoverflies of Northwest Europe* by Mark van Veen has been the 'go-to' text for two decades, but has been superseded by *Hoverflies of Britain and North-West Europe: A Photographic Guide* by Sander Bot and Frank Van de Meutter (2023). *A Naturalist's Handbook* (No 5), by Francis Gilbert, published in 1993, is not intended as an identification guide but provides much interesting information about the natural history of hoverflies. *The Natural History of Hoverflies* by Graham Rotheray and Francis Gilbert, published in 2011, also describes what is known about this family of flies but does not cover identification. This Princeton **WILD**Guides publication therefore plugs a gap and provides an opportunity for people with at least a passing interest in hoverflies to get started. Those who wish to develop their knowledge further need to consult more comprehensive British or European monographs.

Inevitably, a number of technical terms are used when referring to hoverflies. Whilst these are explained in the text where appropriate, for ease of reference most are defined in the **Glossary** (see *page 44*).

Is it a hoverfly?

Class	INSECTA			
Order There are 30 or so orders in the class Insecta, of which *Diptera* and *Hymenoptera* are two	**Diptera** True flies Within Diptera are around 165 extant families – 107 in Britain and Ireland (of which the Syrphidae is one). **Two (one pair) wings:** *See following text*	**Hymenoptera** Sawflies, bees, wasps and ants The Hymenoptera are important in the scope of this book as many British and Irish hoverfly species mimic species within this order. **Four (two pairs) wings:** *See details below*		
Family	**Syrphidae** Hoverflies **vena spuria present:** *See following text*	**Apidae** Bees (including honey-bees, carpenter bees, cuckoo bees and bumblebees)	**Andrenidae** Solitary bees	**Vespidae** Wasps
	There are around 6,000 species of hoverfly worldwide in 200 genera, of which currently 285 species of 69 genera have been recorded in Britain and Ireland. This book covers all the genera and 177 species in detail.	In Britain and Ireland there are ten species of hoverfly that can be regarded as bumblebee mimics, and five or six that can be regarded as honey-bee mimics. Many of the other hoverflies mimic solitary or social wasps and bees in a fairly general way.		

Hoverfly classification

Hoverflies are 'true flies', belonging to the very large insect order **Diptera**. The first question must therefore be "Is it a fly?" The clue to recognising true flies is in the name 'Diptera'. It comes from the Greek, 'di' = two and 'ptera' = wings. The characteristic feature of true flies is that they only have two wings (*i.e.* one pair), as compared with most other insects, including bees and wasps, which have four wings (*i.e.* two pairs). In Diptera, the 'hind pair' of 'wings' have evolved into

A haltere is visible in this image of the hoverfly *Syritta pipiens*.

small, club-shaped appendages called '**halteres**' that help control flight and contribute to these insects being such accomplished fliers.

A true fly also has the following features:
- A body composed of three sections (like all insects): the head, thorax and abdomen.
- The head has a mouth on the underside, a pair of large, compound eyes, a pair of antennae ('feelers') and usually three 'simple eyes' or ocelli placed at the top, between the compound eyes.
- The thorax bears three pairs of legs and just one pair of wings with the halteres located on each side, just behind and below the wing bases.
- An abdomen, consisting of a series of overlapping segments, with the genitalia located at the tip.

Hoverflies form the family **Syrphidae** and, as with all flies, wing venation is often important in their identification. The veins are rod-like struts that support the flexible membrane that forms the wing surface. Most of the veins run out from the wing base towards the wing tip, but there are also **cross-veins** that form braces between them. Hoverflies have an additional 'spurious vein' (**vena spuria**) which cuts across the normal venation and is actually the reinforced hinge of a longitudinal fold in the wing surface. **The presence of the vena spuria immediately tells you that it is a hoverfly.** This feature is more obvious in some species than others, but there is only one British hoverfly (the rare *Psilota anthracina*) that lacks it altogether.

Another feature of the wing venation is the presence of two outer cross-veins, both near to and parallel with the wing margin, which form a '**false margin**'. Few other Diptera have two outer cross-veins (many have just one or none at all), but the other families with this feature are generally bristly flies.

There are a number of other features that can also help to separate hoverflies from other Diptera:
- They are of fairly compact build, without very elongated bodies, very long legs, a long proboscis or extravagantly developed genitalia.
- Many have black-and-yellow, wasp-like markings or are honey-bee or bumblebee-like, although there are also quite a number of black species. Whilst a few of the darker species have bronzy, greenish or even purplish-metallic reflections, the British species are never brightly metallic blue or green like a green-bottle or blue-bottle fly.
- They may be furry (like a bumblebee – see *Mimicry*, page 30), but they are never bristly like house flies or blue-bottles and they do not have bristly legs or strong bristles on the top of the thorax like many flies.
- And, however much they might look like wasps or bees, they are harmless and do not bite or sting!

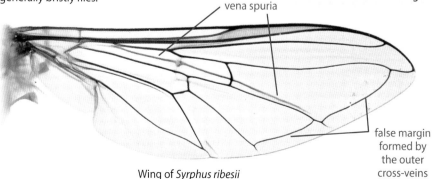

Wing of *Syrphus ribesii* showing the vena spuria and false margin diagnostic of British hoverflies.

The life-cycle of a hoverfly

Eupeodes corollae mating

Episyrphus balteatus laying an egg

Adult
The function of the adult stage is to mate, disperse and lay eggs. Adults may survive from a few days to a few weeks.

Pupa
The pupa is formed inside the skin of the third larval instar and the structure you can see is technically a 'puparium'.

This stage lasts from a few weeks to several months depending on the species.

Hoverfly puparium

Egg
Eggs typically hatch a few days after they are laid.

Egg of *Syrphus*

Larva
The role of the larva is to feed. There are three larval stages or 'instars'. The mature (third instar) larva is the stage at which most species spend the majority of the year and during which most of the feeding is done.

Young *Episyrphus* larva among aphids

Hoverfly biology

Adults

The adult stage is what we normally see and think of as 'the hoverfly'. They are relatively short-lived and survive for a few days to a few weeks. One mark-release-recapture exercise revealed that the large, black-and-white Great Pied Hoverfly *Volucella pellucens* can live for at least 35 days, although the average adult lifespan was 12 days. Recent mark-release-recapture studies of *Hammerschmidtia ferruginea* found the maximum adult lifespan to be 55 days. The primary function of the adult is to mate, disperse and lay eggs. Energy and protein are required to form the eggs. Adult hoverflies usually obtain these by visiting flowers to obtain energy-giving nectar and protein-rich pollen. Most hoverflies lack specialised mouthparts, preferring to visit flowers in which the nectar and pollen are exposed and easy to reach. White umbellifers, such as Hogweed, and members of the daisy family, such as thistles and knapweeds, are favourites.

♂ *Eristalis pertinax*

One instantly recognisable genus of British hoverflies, *Rhingia*, is exceptional, however, in having its mouthparts extended into a long snout, or rostrum. The rostrum has a groove on the underside into which the rather elongated proboscis fits. Consequently, it can visit flowers with a moderately deep tube, like Red Campion and Bluebell, to take advantage of a food source that is out of the reach of other hoverflies.

The closely related genera *Xylota*, *Brachypalpoides* and *Chalcosyrphus* tend not to visit flowers at all. Instead, they feed on honeydew, the sugary secretion of aphids, and pollen grains stuck to the leaf surface.

♂ *Rhingia campestris*

They run over the surface of leaves, rapidly turning backwards and forwards in a very characteristic manner that makes them look like an ichneumon or spider-hunting wasp.

A number of smaller hoverflies, notably members of the tribe Bacchini (*Melanostoma*, *Platycheirus*), visit the flowers of wind pollinated plants such as plantains, grasses and sedges to obtain pollen. These plants provide a particularly rich food source through the production of great quantities of pollen.

Males of many hoverflies show some degree of territorial behaviour. They guard sunny spots by either hovering in a shaft of sunshine or sitting on a sunny leaf, from which they dart out at other passing insects either to chase off a rival or to mate with a female. The same individual will return repeatedly to the same spot. Mark-release-recapture studies have also found that the same male of *Volucella pellucens* will return to hover in the same position along woodland rides on several successive days, indicating a strong and sophisticated level of territoriality.

Female *Eristalis nemorum* on a flower with two males hovering above.

The very noticeable high-pitched whine often heard in British woodlands on sunny summer days is generated by thousands of male *Syrphus* species involved in territorial behaviour. In order to be ready to move at a moment's notice, they keep their wing muscles 'ticking over'. This causes the thorax to vibrate and this vibration is transmitted to the leaf surface, which acts like a sounding board.

Courtship is usually very brief. If the male encounters a female during his darting territorial flights, he grapples with her and, if accepted, they couple, either in flight or after falling onto vegetation. Mating usually lasts for a few minutes.

The small, honey-bee-like hoverfly *Eristalis nemorum* exhibits a more conspicuous courtship-related behaviour than most other hoverflies. A female is often seen feeding on a flower with a male hovering a few inches above her. It seems that this behaviour is linked to the female's readiness to mate but the precise process is not fully understood. If a male has mated with a female, it is not unusual, across a range of insect groups, for him to then stand guard over her to stop other males mating with her before she has a chance to lay the eggs he has fertilised. This may be the case with *E. nemorum*, but occasionally two or more males may be involved, jockeying for position over the female, suggesting that mating has not happened and that the males are waiting in hope. There is a lot more to learn by careful observation!

Occasionally you may encounter a hoverfly with a distended abdomen with distinct pale whitish bands emerging between the segments. Such flies are usually dead and are firmly fixed to a plant; the genera *Melanostoma* and *Platycheirus* are particularly afflicted. These have succumbed to a fungus that grows inside the insect and emerges between the body plates in order to release its spores. Mass occurrences of dead hoverflies attached to the undersides of umbellifer flowers or to grass stems are most frequently encountered in late summer and early autumn. Fungi that attack insects are referred to as 'entomophagous' and are one of the more noticeable causes of death among hoverflies.

Melanostoma killed by the fungus *Entomophthora*.

Eggs

The ovipositor of female hoverflies consists of soft and flexible telescoped sections which, due to a complex arrangement of muscles and sensilla, has incredible sensitivity and manoeuvrability. It enables the fly to deposit eggs, singly or in small batches, very carefully and accurately, in a suitable place for the larvae to find food. Species whose larvae feed on aphids often lay eggs in or near aphid colonies. Those species with plant-feeding larvae usually lay eggs on particular host plants. Identifying the plant on which a female is egg-laying can, therefore, provide a useful clue to the identification of species within some genera (*e.g. Cheilosia, Merodon, Eumerus*).

Hoverfly eggs among aphids.

Hoverfly eggs are usually somewhat elongate-oval in shape with one end wider than the other and often slightly curved. They are typically cream-coloured or pale yellow with a sculptured surface of fine pits and ridges or net-like patterns. These patterns can be distinctive and, potentially,

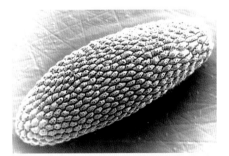

Egg of *Cheilosia vernalis* through a scanning electron microscope.

used to identify a species, but very high magnification is required for them to be fully appreciated. When seen through a scanning electron microscope the patterns are often quite beautiful. Larger species generally lay larger eggs: those of *Volucella* are around 2 mm in length, those of *Syritta* only 0·5 mm.

In the few cases where it has been observed, eggs hatch after quite a short period, typically no more than five days, but this is highly dependent upon temperature.

Larvae

Hoverfly larvae are typical of many dipteran larvae, being maggot-like, without legs or a head capsule. They are often somewhat wider at the rear end and narrow towards the head. Fly larvae usually have two prominent breathing tubes at the rear end: the **posterior spiracles**. In hoverflies, these are fused into a single structure: the **posterior breathing tube**. This is a characteristic of the hoverfly family which enables their larvae to be distinguished from most other fly larvae.

Larvae pass through three stages, or **instars**, during their growth. At the end of each stage they shed the outer skin so that the next stage can grow larger. The first two instars normally last just a few days. However, the third instar can last for weeks, months or even years, depending on the species. In many ways, the third instar larva is the hoverfly. This is the stage that does most of the feeding, and its requirements largely determine the habitat and ecological importance of the species.

The ways in which hoverfly larvae live and feed are amazingly diverse and are reflected in their shapes and specialisations. The larvae of most British flies are poorly known and for some families they are virtually unknown. However, this is not true of hoverflies, since the larvae of about two thirds of the species have been described and the biology of around half is known in some detail. Keys to these species are included in Graham Rotheray's *Colour Guide to Hoverfly Larvae* (1993)*, but there are still plenty of species that require detailed study.

Larva of *Epistrophe* among aphids, showing the characteristic posterior breathing tube.

*Out of print but a PDF can be downloaded from the Dipterists Forum website (see *page 338*).

Aphid-feeding larvae

Hoverfly larvae are well known for feeding on aphids and are consequently regarded as 'gardener's friends'. In reality only about 40% of British species are predators of this type and aphids are not the only prey. A few species feed on a variety of other soft-bodied insects, including coccids and psyllids, and a few feed on other insect larvae. For example, the larvae of *Xanthandrus comtus* feed on gregarious micro-moth caterpillars and those of *Parasyrphus nigritarsis* feed on the larvae of leaf beetles on willows and Alder along streams and rivers.

A hoverfly larva lifts an aphid from the surface of the plant as it feeds.

Predatory hoverfly larvae usually have an extensible and rather mobile front part of the body which they use to probe for prey. They are blind, so rely upon a combination of chemical senses and touch to locate prey. Having located its prey, the hoverfly larva pierces the skin with long thin mouth hooks and sucks out the contents of its victim's body. They have rather sticky saliva which is used to help hold the prey in place whilst they do this. Finally, the empty husk is discarded. As it does this, the larva often rears up and lifts the aphid off the leaf. It is thought that this behaviour helps to prevent alarm pheromones released by the aphid reaching the rest of the colony. It may take a first instar larva several hours to deal with a single aphid, but a fully grown larva can consume an aphid in a minute or two.

One of the most unusual features of predatory larvae is that they are coloured, unlike most other hoverfly larvae, which are pale yellowish-white. Being coloured helps to provide camouflage, which reduces the risk of predation when they are moving around on the surface of a plant in search of prey. Many such larvae are green, often with longitudinal stripes of white, yellowish or brown blotches to break up their outline. Some of the most striking patterns are found in *Dasysyrphus*, the larvae of which feed at night on tree-dwelling aphids and rest on branches and trunks

Larvae of *Dasysyrphus tricinctus* (TOP) and *D. albostriatus* (MIDDLE), camouflaged for resting on bark; and the bird-dropping mimic larva of *Meligramma trianguliferum* (BOTTOM).

Galls on Lombardy Poplar leaf stems caused by *Pemphigus* aphids (LEFT), and the aphids being predated by the larva of *Heringia heringi* (RIGHT).

during the day. They have 'frilly edges' to break up their outline and patterns of reds, greys and browns that camouflage them perfectly against the bark. Another remarkable larva is that of *Meligramma trianguliferum*, which is a perfect bird-dropping mimic.

The larvae of the tribe Pipizini tend to specialise in feeding on aphids inside galls and leaf curls. The larvae of *Heringia heringi*, for example, are common in the galls on the stems of poplar leaves induced by the aphid *Pemphigus*. These larvae have less need of camouflage patterns because they spend much of their time hidden away inside galls – although some pipizine larvae are green.

A predator of beetles

Parasyrphus nigritarsis was once regarded as a great rarity, and even today adults are not reported that frequently. Eggs and larvae, on the other hand, are quite readily detectable. Recent work has shown that *P. nigritarsis* is a regular predator of the larvae of the Green Dock Beetle *Gastrophysa viridula* which lays large clusters of bright orange eggs on the undersides of dock leaves. The hoverfly lays its eggs singly on or adjacent to the egg cluster and its larvae emerge shortly after those of the beetle. Observers with good eyesight can often find the eggs of the hoverfly by diligent searching (mainly across the north and west of Britain). Once the larvae emerge, they too can be found in association with the beetle larvae whose presence is often apparent because they create a lattice of windows in the dock leaves. The hoverfly larva is quite unmistakable and so looking for larvae has become a common way of recording *P. nigritarsis*.

Larvae in the nests of ants

The larvae of the striking black-and-yellow hoverflies of the genera *Chrysotoxum*, *Doros* and *Xanthogramma*, as well as the larvae of *Pipizella*, are believed to feed on ant-attended root-feeding aphids. This has been little studied: for example, it is not known how they get into an ants' nest; whether the female hoverfly enters the nest to lay eggs or whether a larva makes its own way in; and why the ants, which normally protect the aphids they farm from parasites and predators, do not kill the hoverfly larvae.

Doros profuges is listed as a priority species under the UK Biodiversity Action Plan, and a good deal of work has been carried out to try and find out more about it. This work was prompted by a rather obscure 19th-century publication, which suggests that the larvae may be associated with ants nesting in wood, the most likely candidate being *Lasius fuliginosus*. However, the current research has had little success so far in understanding the ecology of this species.

Microdon larvae are highly specialised ant predators which feed on the eggs and brood of the host. The larvae are hemispherical and resemble a woodlouse. Indeed, they were initially classified as molluscs in the 19th

The bizarre hemispherical larva of *Microdon analis* in an ants' nest.

century! The shape is probably an adaptation to protect them from the ants and they are heavily armoured to avoid bites and stings (see *page 298*).

Larvae in the nests of social wasps and bees

Volucella rank among the region's largest hoverflies and, with the exception of *V. inflata*, their larvae live in the nests of bees and wasps. *V. bombylans* larvae usually live in bumblebee nests, and the larvae of *V. inanis*, *V. pellucens* and *V. zonaria* live in the nests of social wasps. The larvae are mainly scavengers in the bottom of the nest cavity, feeding on dead workers and larvae, dropped food and the larvae of other insects that inhabit these nests. *V. inanis*, however, feeds directly on the wasp grubs and its larvae are flattened so that they can fit into a cell in the comb beside their victim. Larvae of *V. pellucens* have also been recorded feeding on moribund wasp larvae in the combs of abandoned wasps' nests late in the season.

Female *Volucella* have to enter the host nest to lay their eggs. Bumblebees sometimes react aggressively and it has been shown that female *V. bombylans* react to being stung by laying their eggs immediately. Observations of *V. pellucens* at nest entrances suggest that the wasps take no notice of them and they can freely enter and leave the nest.

Hoverfly larvae leave the host nest cavity in the early autumn, when the nest is abandoned by the wasps or bumblebees. They pupate nearby: in soil if the nest is underground, or in debris in a tree hole in the case of tree-cavity nests. *V. inanis* quite often use wasp nests in buildings and consequently it is not uncommon for larvae to turn up in houses.

Larva of *Volucella inanis* in a wasp nest.

Plant-feeding larvae

Britain's largest genus of hoverflies, *Cheilosia*, together with *Eumerus*, *Merodon* and *Portevinia*, have larvae that feed in the roots, stems or leaves of plants. *Cheilosia grossa* and *C. albipila* are typical examples whose larvae mine the stems of thistles, particularly Marsh Thistle. Adults of both these species fly very early in the season (late March to April) and lay their eggs on thistle rosettes just as the plant starts to grow. The larvae of *C. grossa* may kill the growing point of the thistle when they start to feed, causing the plant to become multi-stemmed. Mature larvae mine the centre of the stem base, leaving a hollow tube. The larvae are fully grown by July or early August, at which time they exit the thistle by chewing a hole through the side of the stem and pupate in the soil surrounding the plant base.

Adults of *Cheilosia grossa* and *C. albipila* are often overlooked because they fly so early in the year. It is, therefore, much easier to record these species by searching for their larvae in thistle stems (see *opposite*). Mature larvae are relatively easy to tell apart because their posterior breathing tubes have characteristic shapes. Unlike adults, finding larvae is not dependent upon good weather, so a wet day in July can profitably be spent splitting thistle stems! This technique has shown that these species are much commoner and more widespread than records of adults suggest, especially in the uplands.

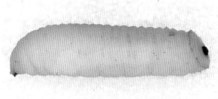

Portevinia maculata larva (TOP), Ramsons flowers (LEFT) and an adult male (ABOVE) sitting on a leaf of the plant.

HOVERFLY BIOLOGY

Recording *Cheilosia albipila* and *C. grossa* by searching for larvae in thistles

The larvae of both species feed in the base of the stem of Marsh Thistle and those of *C. grossa* in other thistles such as Spear Thistle (but seemingly not in Creeping Thistle). They are easily found during May, June and July, or even August in the north, by splitting the base of thistle stems and looking for those that are hollowed out. Experience has shown that both species can be found in the same patch of thistles.

Opening up the rosette of a Spear Thistle.

A good, hefty boot applied to the base of the stem will knock the plant over and open up the rosette. Insert a suitable implement and split the rosette and the base of the stem apart.

A clasp knife will do the job, but a small-bladed, sturdy gardening trowel or fork is probably better. It should be immediately obvious whether the plant has been tunnelled. An unaffected stem is filled with green, solid pith; a tunnelled stem is hollow and usually quite stained within by obvious deposits of brown frass. When you find a tunnel, split the stem upwards until you find the larva. Late in the season, you may find

A *Cheilosia albipila* larva *in situ* in its tunnel after the stem has been split.

that the tunnel is no longer occupied – in which case it is worth a quick dig around the roots to look for mature larvae or puparia.

The larvae of these two species are quite easy to tell apart by the characteristic shapes of the posterior breathing tube (PBT). *Cheilosia albipila* is white and the PBT is parallel-sided and has a 'dagger point' projecting between the tips of the two fused hind spiracles. *Cheilosia grossa* is a dirty brown colour and the PBT is broad and blunt with a flange either side.

Other species of *Cheilosia* that occur in thistles are generally similar, but lack the diagnostic 'dagger point' or 'flanges'. Larvae of *C. fraterna* and *C. proxima* favour the more slender growth and side branches higher up on the plant, but little is known about others such as *C. mutabilis* and *C. cynocephala* (the latter of which is thought to be restricted to Musk Thistle). You may also find other insect larvae, such as caterpillars of the Frosted Orange moth *Gortyna flavago*.

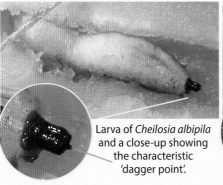

Larva of *Cheilosia albipila* and a close-up showing the characteristic 'dagger point'.

Larva of *Cheilosia grossa* and a close-up showing the characteristic flanges on the broad, blunt PBT.

Cheilosia semifasciata larva in a leaf-mine on a Navelwort leaf (TOP LEFT); the plant growing in profusion on a slate rock face in north Wales (RIGHT) and an adult female of the fly (BOTTOM LEFT).

The common *Cheilosia albitarsis* and its close relative *C. ranunculi* feed on the roots of buttercups. In these cases the larvae live on the outside of the root.

Cheilosia semifasciata is relatively unusual for a hoverfly in that its larvae are leaf-miners in Navelwort in north Wales and Orpine in southern England. Since a single leaf does not provide sufficient food for a larva, they have to move from one leaf to another several times during their growth. They make quite conspicuous blotch mines, especially once the leaf starts to die, and as these blotches would be easily noticed by hungry birds, the larva chews through the petiole so that the leaf falls off as it exits the mine.

The larvae of *Merodon* and *Eumerus* mine garden bulbs, especially daffodils. It seems likely that they need the bulb to be already damaged in order to gain entry and that they do not feed directly on the plant material, but on the fungi that rot the damaged bulb.

Cheilosia longula and *C. scutellata* larvae live in the caps of *Boletus* and some other large fungi. Up to 50 larvae have been recorded in a single *Boletus* cap (they are, after all, among the largest mushrooms!). As the larvae consume the mushroom, it is reduced to a dark brown patch of slime and the mature larvae pupate in the soil underneath.

Saprophagous larvae

Saprophagy involves feeding on dead or decaying organic matter and is the commonest way of life for hoverfly larvae. Saprophagous larvae are filter-feeders, sucking in micro-organisms from fluids, and consequently are found in moist or wet situations. Different genera favour particular micro-habitats, as detailed in the following paragraphs.

Sap runs

Sap runs on trees occur when the tree has suffered some form of damage. Such runs may be transient if the bark manages to seal itself up again, or may persist for many years, especially if micro-organisms such as bacteria become established. Some are big and obvious, but most are quite small and hard to detect. Whilst all trees may develop sap runs, some seem to be particularly prone, notably Horse-chestnut, elms and Yew. Sap runs are an entirely natural phenomenon and not something to worry about in terms of the health of the tree. They should not be used as an excuse to cut down 'afflicted' trees!

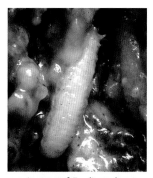

Larvae of *Ferdinandea* in a sap run.

Sap contains lots of nutrients and provides an ideal medium for yeasts and bacteria to thrive. Consequently, sap runs often develop a yeasty, brewing smell. Although the larvae of quite a few hoverflies can be found feeding on micro-organisms in the sap in these runs, *Brachyopa* and *Ferdinandea* are the most frequent. The larvae of these species are somewhat flattened and have projections at the side of the body which give them a spiky look. Larvae can occur in considerable numbers and it is not unusual to find a mixture of sizes, suggesting that development may take more than one year. *Brachyopa* larvae seem to have an amazing ability to resist desiccation when sap runs dry out, which they tend to do from time to time.

Rotting wood

A number of hoverfly larvae live in rotting wood. Decaying stumps and roots are favoured, particularly where the rotten wood is soft and wet and has developed a porridge-like consistency. Secondary decay by micro-organisms follows the initial decay, which is usually by fungi and can penetrate remarkable distances into the wood. Hoverflies such as *Brachypalpoides lentus* and the genera *Criorhina* and *Xylota* are among the commoner inhabitants of rotten wood, but some of the deadwood rarities, such as *Caliprobola speciosa* and *Blera fallax*, may also be found.

Bumblebee-mimicking females of *Criorhina* can often be seen flying around the bases, or inside hollows, of old stumps or crawling into crevices between large roots, looking for egg-laying sites. This would be unusual behaviour for a bumblebee (unless there is an underground nest in the stump, in which case a stream of them will be going in and out of the same spot). So, if you see 'bumblebees' behaving like this they are worth checking!

 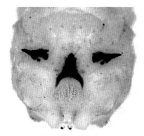

Larva of *Criorhina ranunculi* showing the large hooks developed on the thorax. These are used by the larva to get around in rotting wood and lever firmer splinters out of the way.

Rot-holes

When a branch breaks off from a tree, fungi may become established in the broken stump, and, over time, a rot-hole develops. The holes in some trees, such as oaks and Ash, develop a dry rot, but in others, such as Beech, Horse-chestnut and Sycamore a wet rot develops. If the hole is positioned such that it can collect water as well as falling twigs and leaves, a layer of wet decaying material gradually builds up. Wet rot-holes provide ideal conditions for a range of hoverfly larvae, but are especially favoured by species within the genera *Brachypalpus*, *Callicera*, *Mallota*, *Myathropa*, *Myolepta*, *Pocota* and *Xylota*.

The micro-organisms responsible for breaking down the decaying material use up the available oxygen, leaving a substrate which is often black and very smelly. This lack of oxygen makes it difficult for hoverfly larvae to breathe, and one of the adaptations seen in many of the larvae that live in such conditions is the elongation of the end of the body and the posterior breathing tube to form a telescopic snorkel which allows the larva to reach the air. Larvae in the genera *Caliprobola*, *Myolepta* and *Pocota* are described as 'short tailed', as the body is only moderately extended, whereas in *Mallota* and *Myathropa* the breathing tube is longer than the larva's body and consequently they are known as 'rat-tailed maggots'.

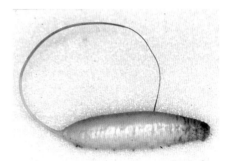

Myathropa florea larva (TOP);
a rot-hole in a lime tree (BOTTOM).

In some cases it may be easier to find the larvae than the adults and consequently some species are easier to record by searching suitable looking rot-holes. *Callicera rufa*, for example, was once regarded as a very rare species of Caledonian pine woods as the adults had only ever been found a handful of times (see *page 43*). However, the Malloch Society (an organisation of Scottish dipterists) discovered that larvae could be found by searching rot-holes in Scots Pine trees, and found them to be widespread in pine woods in northern and central Scotland and occasionally in plantations of other conifers, such as Larch. This approach is potentially applicable to other seemingly rare rot-hole breeders.

Larvae living in rotting wood and rot-holes often take more than one year to develop. *Callicera* larvae (see *page 43*) typically take two years and may sometimes take as long as four or five. Consequently, it is not unusual to find very different sized larvae living together in the same cavity.

HOVERFLY BIOLOGY

An Aspen log inhabited by *Hammerschmidtia ferruginea* (LEFT); adult male (TOP RIGHT) and larva (BOTTOM RIGHT).

Under bark
When a tree falls, or a large branch breaks off, the sap under the bark decays and forms a wet, pungent layer which is full of food for hoverfly larvae. This is only a temporary condition because once the bark starts to crack and peel off it dries out. Logs therefore only provide suitable conditions for a few years. Hoverfly larvae found in this situation include the genera *Chalcosyrphus*, *Hammerschmidtia* and *Sphegina*. *Sphegina* favour logs in wet places, such as wet woodlands whilst *Chalcosyrphus eunotus* is particularly associated with partially submerged logs in wooded streams.

Other wet situations
A wide range of hoverfly larvae are associated with accumulations of wet, rotting vegetation of one sort or another, including compost heaps, dung and manure piles, and accumulations of decaying material in ponds and ditches. These larvae filter-feed on the micro-organisms responsible for the decay. Some of the larvae that live underwater are of the 'rat-tailed maggot' type (*Eristalis*, *Eristalinus*, *Anasimyia*, *Helophilus*, *Lejops*, *Parhelophilus*) and are often found in the decaying vegetation in the bottom of reedbeds and similar situations. *Eristalis tenax* is the archetypal 'rat-tailed maggot'; its breathing tube can be several times the length

 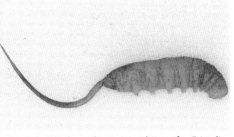

A corner of a small garden pond full of rotting leaves – ideal egg-laying conditions for *Eristalis* and *Helophilus* (LEFT); *Eristalis tenax* larva (RIGHT).

of the body when fully extended. *Sericomyia* also have long-tailed larvae, which are found in more acid waters, such as in peat bogs. *Eristalinus aeneus* is found in pools near the sea-shore where decaying seaweed accumulates, although in other parts of Europe it is not a coastal species and is found in manure heaps and farmyard slurry pits.

The larvae of some genera within the tribe Chrysogastrini (*e.g. Chrysogaster* and *Melanogaster*) employ a highly adapted strategy for breathing. Their posterior breathing tube is a stout spike which they use to penetrate the stems and roots of emergent plants to reach the air passages within, enabling them to sit and filter-feed underwater, out of harm's way.

Rotting vegetation

Some of the more terrestrial species, such as those in the genera *Syritta* and *Neoascia*, live in piles of wet decaying material including compost, farmyard manure or silage.

Rhingia campestris larvae live in cow dung. The female attaches her eggs to grass blades above cow-pats, and when the larvae hatch they drop onto the surface and burrow in. The cow-pat needs to be fairly fresh because once it has developed a crust the larvae may not be able to penetrate it. However, as this species is abundant in parts of the country where cows are scarce or absent (such as in woodlands in arable areas of the East Midlands and the Fens), it must be able to utilise larval media other than cow-pats.

Pupae

The pupa is formed within the larval skin of the final instar so, technically, what you see is a '**puparium**' with the actual pupa hidden inside. The pupal stage may last from a week or two to several months, depending on the species. Hoverfly larvae usually pupate near the larval habitat and, if the pupa is the over-wintering stage, it is generally well hidden in soil or leaf-litter.

Migration

Migration in insects is somewhat different from the behaviour seen in migratory birds insofar as it is not the same individual that travels north in spring to breed and returns to warmer climes for the winter. Many insects, such as Red Admiral and Clouded Yellow butterflies, arrive and breed locally in Britain and Ireland, but it is their progeny that return south (although it is an open question as to how many successfully complete the return trip).

There are a few hoverflies that migrate and do not seem to be permanently resident in Britain and Ireland. One of the commonest is *Scaeva pyrastri*, which is frequent in some years but hardly seen at all in others. Most other migrants are species that are normally resident, but in some years they are augmented by immigration from continental Europe. Hoverflies that are well known for this include *Episyrphus balteatus*, *Syrphus vitripennis*, *Eupeodes corollae*, *E. luniger*, *Sphaerophoria scripta*, *Helophilus trivittatus* and various *Eristalis* species. They are sometimes encountered arriving in abundance on the south or east coasts. Occurrences of large numbers of black-and-yellow-striped hoverflies turning up on bathing beaches have led to press reports of 'plagues of wasps'.

Migratory species are capable of crossing considerable stretches of ocean. Evidence of this comes from reports from North Sea oil platforms where hoverflies are sometimes seen in vast numbers. More recently, studies in the Mediterranean suggest that there is a massive migration from Asia into Europe in the spring and radar data have also highlighted phenomenal movements of hoverflies in some years.

Some hoverfly species have only been recorded very occasionally in Britain and are regarded as 'vagrants'. They have often occurred in locations that are also well known for their vagrant birds, such as Fair Isle, North Norfolk and along the south coast. Examples of such species are *Scaeva albomaculata*, *Helophilus affinis* and *Eupeodes lundbecki*.

Didea alneti, which can be extremely abundant in northern Europe, is an interesting case. It is very rare in Britain and Ireland and can occur at a locality, sometimes breeding for a year or two, before dying out. The last known occurrence was in a conifer plantation in the south of Northumberland in the late 1980s. The species is believed to be irruptive, only arriving in Britain following a particularly good breeding season in Scandinavia. It has unusual abdominal markings which are often green or turquoise-blue rather than yellow.

♂ *Sphaerophoria scripta*

HOVERFLY BIOLOGY See also *Identifying wasp and bee mimics pp. 62–65*

Polymorphism and other colour variations

A number of hoverfly species show markedly different colour forms (polymorphism) within their population. This is most obvious among species that mimic bumblebees. *Merodon equestris* (*p. 248*), *Volucella bombylans* (*p. 270*) and *Criorhina berberina* (*p. 294*), for example, all have two or more colour forms which mimic the patterns of different groups of bumblebees. *Merodon equestris* is perhaps the most extreme example, with at least four (and, according to some authors, anything up to seven!) named varieties, but many individuals have intermediate characters, and it is therefore difficult to name them consistently.

The sexes of some species, such as *Criorhina ranunculi* (*p. 292*), *Eristalis intricaria* (*p. 228*) and *Leucozona lucorum* (*p. 116*), have different colour patterns (*i.e.* they are sexually dimorphic), and there are others where there are morphological differences between the sexes (*e.g. Cheilosia pagana*, where the orange antennae are much bigger and more obvious in females than in males – see *page 181*).

'Brood dimorphism' is a difference in appearance between the spring and summer broods of a species. This is particularly prevalent in those species of *Cheilosia* that have two broods per year; in this instance, the spring brood is typically larger and longer haired than the summer brood.

Many yellow-and-black/brown hoverflies are very variable in the extent and brightness of their markings. This has been found to be related to the temperature at which they were reared. For example, in the laboratory *Episyrphus balteatus* larvae reared in cool conditions produce dark-coloured adults, whereas those raised at higher temperatures are brightly coloured and with reduced black markings. This tendency is also evident in the field: adult *E. balteatus* emerging early in the year (having developed during the cooler winter months) tend to be much darker in colour (*p. 163, bottom*) whilst midsummer individuals tend to be brightly coloured (*p. 163, top*). This is believed to aid temperature regulation because a basking hoverfly with more dark pigment will absorb sunshine better and warm up more quickly.

Many genera that breed throughout the year, such as *Eupeodes*, *Meliscaeva*, *Eristalis* and *Myathropa*, show similar variability and, unfortunately, this can make them more difficult to identify.

Variation in the markings of *Myathropa florea*: Compared with a typical individual from mid-May (LEFT), this one from August (RIGHT) has noticeably brighter and more distinct markings.

Some of the many colour forms of *Merodon equestris*

Form *validus* – thorax completely dark; 'tail' greyish or buff.

Form *narcissi* – both thorax and abdomen tawny.

Form *transversalis* – like *narcissi* but with a broad black band across the abdomen.

Another, rather attractive, example of *narcissi*, with an orange thorax and pale abdomen.

Form *equestris* – thorax yellow at the front; 'tail' greyish or buff.

Form *bulborum* – like *equestris* but with a red/orange 'tail'.

Mimicry

Many hoverflies resemble bees or wasps, not only in their shape and colouring, but also in their behaviour. Some are extremely good mimics and even experienced recorders need to look twice. Others are not nearly such convincing mimics; for example, many black-and-yellow Syrphini only vaguely resemble wasps, whilst several *Eristalis* species have only a passing resemblance to honey-bees – see table on *page 63*.

As mentioned previously, *Volucella bombylans* is a very convincing bumblebee mimic that breeds in bumblebee nests. Is it trying to fool the bees in order to gain entrance to the nest to lay eggs? This seems unlikely. The colour forms of adults bred out from larvae and pupae found in nests of a particular bumblebee are not the same. The probability that the colour form of the *Volucella* that emerges matches the host bee is no better than random chance. Its close relative, *V. pellucens*, which breeds in wasps' nests, looks nothing like a wasp; it has a black body with large white markings at the base of the abdomen, though observations suggest that it has no problem entering a nest.

If they are not trying to fool the bees and wasps, who or what are they trying to convince? The usual explanation is **Batesian mimicry**, a hypothesis first proposed by the English naturalist H.W. Bates. The theory is that a palatable species evolves colour patterns which imitate the warning signals of a noxious species, directed at a common predator. In other words, by looking like a stinging and distasteful bee or wasp, a hoverfly gains a degree of protection from predators, even though it is actually perfectly edible.

Experiments with birds such as starlings and pigeons suggest that they have abilities similar to humans in distinguishing between hoverflies and wasps or bees: *i.e.* they are just as likely to be confused and may have to look twice! Observations of a confiding Spotted Flycatcher showed that it could tell the difference: it would carefully remove the stings of wasps before eating them, but showed no tendency to try and do the same with a wasp-mimic hoverfly.

However, there is little evidence that birds are significant predators of adult hoverflies. The predators that are most regularly seen feeding on them are other invertebrates, especially spiders, dragonflies, wasps and other flies, such as females of the Common Yellow Dung-fly *Scathophaga stercoraria*. Indeed, the solitary wasp *Ectemnius cavifrons* specifically targets hoverflies to stock its nest.

Two hoverfly predators: *Scathophaga stercoraria* (LEFT) and *Ectemnius* sp. (RIGHT).

HOVERFLY BIOLOGY

The Drone Fly *Eristalis tenax* (LEFT) and a honey-bee (RIGHT).

Volucella bombylans (LEFT) and a bumblebee (RIGHT).

Chrysotoxum cautum (LEFT) and a Common Wasp (RIGHT).

Finding hoverflies

Hoverflies are found throughout Britain and Ireland, although more species tend to occur in the south and east and fewer in the north and west. The northern species do include some of the region's most charismatic hoverflies such as *Blera fallax*, a species of the Boreal forest that maintains a tenuous foothold in central Scotland. Northern species may occur farther south on higher ground or in habitats more typical of northern areas, such as conifer plantations.

Climate is one important factor that influences this distribution. Hoverflies tend to prefer the warmer and drier, continental conditions of south-east England than the wetter and cooler Atlantic conditions of northern and western areas. Some southern species have a more coastal distribution towards the north of their range where the influence of the Gulf Stream ameliorates the local climate.

Geology also plays a part; the underlying rocks of southern and eastern Britain are younger, less acidic and provide richer soils that host a greater biodiversity than the older, more acidic rocks of northern and western Britain.

Land-use is also a significant factor. This can be seen in the distribution maps of many species where there is a dearth of records from some intensively farmed areas such as Lincolnshire and the East Anglian Fens. In these areas, hoverflies can be hard to find and are largely restricted to surviving habitat fragments such as roadside verges, ditches and churchyards. Another example is conifer plantations, which tend to be located in northern and western areas and, although inhabited by a good number of species, they support considerably fewer compared with extensively wooded areas of the south and east that have retained continuous tree cover over the centuries. Unsurprisingly, these areas support the region's richest diversity of hoverflies.

Species lists from a good site in southern England may be in excess of 100 species (the richest sites have nearer 150 species). Farther north, numbers drop quite dramatically and, in the East Midlands for example, a good site may support around 90 species. The farther north you go, the more numbers of species tend to drop, so that sites in the highlands and islands of the far north and west may support only around 30 species.

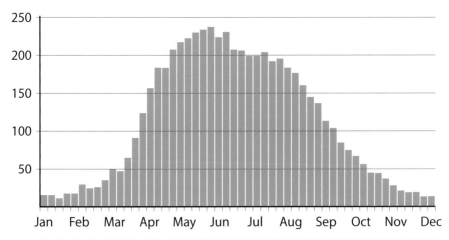

Number of species recorded each week from 2010 to 2021 by the Hoverfly Recording Scheme.

The seasonal calendar

Many hoverflies are easy to find at flowers and you will quite rapidly develop a decent list of species that visit common grassland and hedgerow flowers. However, there are many other places you should look if you wish to extend your list. Fieldcraft is something that develops over time, and is as much about knowing what you are looking for as it is about employing general observational skills.

It is also worth bearing in mind, that as the season progresses, different search techniques can be employed to find some of the species that are not natural flower visitors. Once you start to develop a search pattern, you will find yourself looking out for specific features such as a sunlit tree trunk or highly reflective young leaves in a sunny spot. However, the best way to starting recording hoverflies is to follow the season and the flowering plants that attract them.

You can look for hoverflies throughout the year, even in midwinter when a few hardy species such as *Episyrphus balteatus*, *Eupeodes luniger*, *Meliscaeva auricollis* and *Eristalis tenax* will visit garden flowers such as crocuses, Winter Jasmine and *Viburnum × bodnantense*. A sunny day in January will often yield results; the first couple of records of this millennium were of *Eristalis tenax* on 1st January 2000!

The field season really starts in **March** once Goat Willow, Blackthorn and Cherry Plum come into flower. Sometimes these flowers can be alive with flies, making it difficult to pick out a hoverfly from a sea of white petals or to catch one from among interlocking branches. It is therefore often a good idea to focus on a single branch that can be viewed in isolation against a neutral background or the sky. Scanning rides for hovering flies can also pay dividends as there are several species that defend sunlit spots.

Cheilosia bergenstammi on Dandelion.

Eristalis intricaria on willow (sallow) blossom.

Dasysyrphus venustus on Lesser Celandine.

The commonest is *Eristalis pertinax* whose males often hover at head height or within ready net height. Others such as *Eristalis intricaria* and *Cheilosia grossa* generally fly much higher, so a telescopic net handle becomes invaluable (see *page 318*). Do not assume that every hoverfly you see is the same; eyes can play tricks and you stand more chance of finding new species by checking lots of individuals.

Low-growing spring flowers such as Lesser Celandine and Wood Anemone can be useful lures, but are often rather disappointing. Even so, they will attract commoner species such as *Platycheirus albimanus*, *Cheilosia pagana* and *Melanostoma scalare*. Occasionally a more interesting species will be seen, such as *Melangyna lasiophthalma*. Young foliage, such as the fresh leaves of Cow Parsley, often provides sunny places for insects to sunbathe. This is an excellent way of finding hoverflies and will often yield a good selection of *Helophilus* and *Eupeodes* species. Do not overlook sunlit dry leaves and grass tussocks as some hoverflies will sunbathe on these too.

All of these situations can be checked as you look for a suitable 'honey pot' such as a flower-laden Goat Willow or Grey Willow (later in the spring, Eared Willow in the uplands can be really good). These, and various members of the genus *Prunus*, are the most likely plants on which to find many of the more interesting spring species, such as the bumblebee mimic *Criorhina ranunculi* and various *Melangyna*, *Parasyrphus* and spring-flying *Cheilosia*.

As spring progresses the opportunities for finding hoverflies greatly increase as more and more plants that will attract hoverflies come into flower. These include Hedge Mustard, Wild Garlic, Dandelion, Cow Parsley, Ground-ivy, Bugle and White Dead-nettle. Each is worth a look and may attract a different range of hoverflies. For example, Ground-ivy will attract the long-tongued *Rhingia campestris*, but is also a great place to look for *Heringia* and *Neocnemodon*, which mimic tiny solitary bees as they flit among the flowers. Cow Parsley is a common plant that, at first sight, seems to attract relatively few hoverflies. However, this can often be an illusion because the hoverflies are widely dispersed and a good list of *Cheilosia*, Pipizini and Syrphini may be assembled from a large area of Cow Parsley.

♂ *Cheilosia pagana*

May is an exciting time because diligent fieldcraft can often reveal species that many people miss. This is especially true of the genus *Brachyopa*, species of which do not generally visit flowers and require more specialised search techniques. Finding *Brachyopa* is an art, but once mastered it will be found that species within this genus are by no means as scarce as past texts have suggested. Two species, *Brachyopa scutellaris* and *B. insensilis*, are actually relatively common and both can be found on sunlit tree trunks, close to the ground. *B. scutellaris* is associated with a wide range of trees, including oaks, Ash, limes and poplars, whilst *B. insensilis* is usually found close to Horse-chestnut. *Brachyopa* are also inveterate leaf-baskers and will often be found on sunlit leaves of Sycamore and limes. Although Sycamore is often regarded as an unwanted invader, it is a really valuable tree for the hoverfly hunter.

♂ *Epistrophe elegans*

Many other hoverflies sunbathe and there will be some days when a few species are super-abundant. For example, there is usually a two or three day period when the beautiful *Epistrophe elegans* is found in huge numbers, often with swarms of males around a particular bush.

A sunlit tree base and nearby vegetation provide an ideal location for *Brachyopa*.

From **late April** through to **mid-May** in southern England a wonderful array of hoverflies can be found. A warm, sunny, flowery woodland ride is a delight that is difficult to match. Farther north, although the season is more compressed, finding hoverflies is still just a matter of finding the right flowers and judging whether the day is warm and calm enough to generate hoverfly activity.

♂ *Criorhina floccosa*

Mid- to **late May** in southern England is a time of plenty and the hedgerows and woodland edges are a mass of Hawthorn flowers. These are great lures for hoverflies but really hard to work – there is so much blossom and the quick flitting hoverflies are widely dispersed. This is the time to look for species such as *Criorhina asilica* and *C. floccosa*, as well as more unusual species such as *Brachypalpus laphriformis*. Hawthorn flowers (probably a fortnight or so later) are also good lures in northern England and Scotland. However, there is possibly a better one: Rowan. The flowers of Rowan can be an immensely effective lure for a wide range of Syrphini and will also attract bigger species such as *Sericomyia silentis* and especially *S. lappona*. Rowans are also good trees to search for *Sphegina*, including the brightly coloured forms of *S. sibirica*. You will need a long net handle here too!

Once the Cow Parsley dies away and Hogweed comes into flower the summer is upon us, and the associated hoverfly fauna is there to be found. However, although Hogweed was

Many hoverflies, mainly *Eupeodes corollae* with *Scaeva pyrastri*, *Episyrphus balteatus* and a ♀ *Syrphus* sp., feeding on Hogweed.

once a fantastic lure in southern England it is rarely productive today. The reason for this is not fully understood, but it is noteworthy that Hogweed is still a great lure in wetter parts of northern and western Britain. It may be that recent drier conditions makes the plant less nectar-rich and therefore less attractive to hoverflies. Alternatively, perhaps there are just fewer hoverflies. Recent analysis of trends in occurrence suggests that declines have been particularly pronounced in south-east England. Whatever the reason, it is worth keeping an eye open for other lures such as Upright Hedge-parsley and flowering shrubs such as Dogwood and Wild Privet, both of which can yield many hoverflies, especially species such as *Volucella bombylans*, *V. inflata* and *V. pellucens* together with all of the *Criorhina* species.

May and **June** are when the greatest number of hoverfly species can be found. A really warm, sunny day (but not a dreadfully hot one) will yield 30 or more species from a good site. When looking, time of day is an important factor. The number of hoverflies peaks late morning, dropping off around mid-day during hot weather, before picking up again in late afternoon. Good numbers can often be found in late evening sunshine, making the most of the sun's last few rays – so all may not be lost if you cannot get out of work until late. Surveying hoverflies at this time may also yield useful information because there is comparatively little known about hoverfly activity at this time of day.

It is also worth bearing in mind that, although some hoverflies may appear to be very rare, this could be due to the fact that they have only a short emergence period restricted to just a few days each year. This seems to be the case with *Brachyopa* since when you get a productive day you will often find *B. scutellaris* at every site you visit. The same seems to hold for *Myolepta dubia* which, in Surrey, was found at four disparate locations on the same day, and yet was otherwise very scarce. It is therefore worth making an effort if you think you have hit a period of peak emergence.

The months of **July** and **August** are great times for recording hoverflies as this is when they are at their most plentiful. Resident populations are augmented by migrations of common species such as *Episyrphus balteatus*, *Syrphus vitripennis* and *Scaeva pyrastri*. There are also lots of flowers for them to visit, Wild Angelica, ragworts, Water Mint and Common Fleabane being particularly favoured. For these reasons, if a rich site is worked assiduously a species list well in excess of 30 can be assembled, providing you are able to check the identity of the *Cheilosia* and *Sphaerophoria*, which are often very numerous at this time of year.

Hoverfly numbers drop remarkably rapidly by the end of the third week in **August**, but they can still be plentiful well into **September** and **October**. The number of species recorded by the Recording Scheme for this time of year is not greatly lower than for mid-August, but the number of individuals is considerably less. Although the flowers they like are generally dying off and there are fewer nectar sources, Ivy can be very productive in the autumn.

♀ *Myathropa florea* on Ivy flowers.

Where to look for hoverflies

Hoverflies can be found almost anywhere, from urban gardens through to the seaside esplanade, and there will be occasional surprises. Choosing a site to find hoverflies also involves an element of fieldcraft. You need to know the sorts of habitats particular species favour in order to find them, so it really is important to know something about the animals you wish to find.

Sweeping is an essential part of the surveying process because it often yields species that might otherwise be overlooked. For example, a patch of Creeping Buttercup flowers will often yield more than just *Cheilosia albitarsis*. Here, *Lejogaster metallina*, *Melanogaster hirtella* and perhaps even *Pipiza* species may also be found. *Sphaerophoria* are best sought by sweeping on heathlands, and with a bit of effort several of the scarcer species can be found.

Many hoverflies occupy quite specialised niches. This makes them very useful indicators for conservation management and also means that they must be looked for in particular places. The best sites usually have large areas of high-quality habitats, but this is not an absolute rule and many important records come from small and otherwise insignificant sites. This is exemplified by many of the hoverflies considered to be old woodland indicators: the 'deadwood fauna'. Whilst top quality deadwood sites such as the New Forest, Windsor Great Park, Epping Forest and Burnham Beeches may yield a sizeable assemblage of species from rot-holes and decaying stumps and roots, many if not most of the species involved can be found at much smaller sites. For example, Old Sulehay Forest in Northamptonshire comprises only 35 hectares of woodland that supports very few over-mature trees and yet has yielded an exceptional list of deadwood hoverflies, including *Callicera aurata*, *Mallota cimbiciformis*, *Myolepta dubia* and *Xylota xanthocnema*.

Deadwood hoverflies are not just confined to woodlands. For example, there are two records of *Callicera aurata* from a garden in Wolverhampton. Other great surprises were *Mallota*

Beech stumps in Mark Ash Wood, New Forest.

cimbiciformis in open grassland at Thrislington Plantation near Durham, and *Myolepta dubia* found in open countryside in Surrey adjacent to a spinney of no more than a hectare in extent. In the case of deadwood hoverflies you must expect the unexpected – *Callicera rufa* being a prime example (see *page 43*).

Many other habitats support a range of hoverflies, although some species are more faithful to some habitats (such as heathlands, dry, hot grasslands, and wetlands) than others.

Although heathlands are normally associated with southern localities, the main component of heathland, Heather, is much more widely distributed. This means that whilst some heathland hoverflies such as *Pelecocera tricincta* are restricted to the truly southern heaths of Sussex, Surrey, Hampshire, Dorset and Devon, many others are more widespread. This is particularly true of species in the genus *Sphaerophoria*, such as *S. virgata*, *S. fatarum* and *S. philanthus*, which occur north into Scotland. Some, such as *Trichopsomyia flavitarsis*, are found in acid grasslands as well as on heathland, indicating that the association is more complex than just a relationship with Heather heaths.

Old Beech at High Standing Hill, Windsor Great Park.

Many other apparent heathland associates are actually more closely aligned with the pines that often invade heathland. These habitats can therefore be great places to record species whose larvae are predacious upon conifer aphids. *Didea*, *Parasyrphus* and *Scaeva selenitica* are good examples, though they can also be readily found in conifer plantations.

Studying hoverflies on heathland is a special art and can be hard work, particularly as there are few nectaring sites apart from the Heather itself. The best sites are often those dissected by footpaths with a heathery verge containing yellow composites and Tormentil. Although such verges are often the most productive places to look, in late summer many hoverflies will also visit Heather flowers. Some species are exceptionally difficult to find as adults. *Microdon analis* and the complex of *M. myrmicae/mutabilis* are noteworthy in this respect, but once their larval stages are known they are much easier to find. In the case of *M. myrmicae/mutabilis*, the two species can only be separated on larval and pupal characters and this requires careful examination under high magnification.

Dry, hot grasslands often exhibit similar characteristics to dry heathland; they are relatively warm and are favoured by hoverflies that prefer such conditions. Given the close parallels with heathlands, there is unsurprisingly a great deal of overlap between the faunas. Calcareous grasslands on chalk and limestone do, however, support a suite of relatively specialised species, such as *Microdon devius*, whose larvae are associated with the hill-forming Yellow Meadow Ant *Lasius flavus*. This association is peculiar because although most sites are on chalk downlands, there are sites in north Wales and East Anglia that are much damper and cooler. Quite why this species is not more widespread is somewhat of a mystery, especially as *Xanthogramma citrofasciatum*, which is also associated with the yellow ants' nests, occurs more widely on grass heath and even on grazing marshes.

Experiences from southern Scotland show that if conditions are good and there are plenty of flowers, then hoverflies can occur in good numbers. Devil's-bit Scabious has been found to be particularly popular; in northern and western regions of Britain and in Ireland, areas with this species are good places to search for hoverflies such as *Eriozona syrphoides*, *Didea fasciata* and *Sericomyia superbiens*. This also illustrates the difference in habitat between acidic species-poor localities that support Heather and Devil's-bit Scabious compared with richer soils, which are often much more agricultural.

Finding hoverflies is a much more demanding process where tall flowering plants are not immediately obvious. For example, moorlands can be seemingly barren until you find a damp streamside with Tormentil, which will yield all sorts of small hoverflies, such as *Melanostoma*, *Platycheirus*, *Melanogaster*, *Lejogaster* and *Trichopsomyia*. The Heather itself can be very attractive to hoverflies when it is in flower and, in particular, has been found to support large numbers of *Didea*. So too can Heath Bedstraw, which in Scotland is often an exceptionally good lure in woodland rides. The trick is to try to find flowers that are attracting hoverflies and then develop your technique from there. In some places you may have to sweep the vegetation and here there are definite benefits from retaining a number of specimens: they may look very similar in the net but microscopic examination can reveal several species.

A path on Chobham Common in Surrey. Open ground like this is important on heathland in providing warm areas. Flowers like Tormentil grow along the sides of such paths and are popular with hoverflies such as *Paragus* (INSET).

Wetland hoverflies show similar variation in habitat affinities, with some favouring much more restricted habitats than others. For example, *Anasimyia interpuncta* seems to favour grazing marshes and some riverine wetlands, whereas its near relatives *A. contracta* and *A. lineata* are much more widespread, occurring in a wide range of locations where Bulrush grows, even in small roadside ditches. Some wetland species are even more specialised. *Tropidia scita*, which is readily recognised by the obvious triangular flange on its hind femora, is very closely associated with Common Reed and is most frequently found in coastal locations. Another example is *Lejops vittatus*, which is restricted to brackish grazing marsh colonised by Sea Club-rush and is usually found around the head of tidal estuaries.

Sericomyia silentis on Devil's-bit Scabious.

Many members of the tribe Eristalini (wetland hoverflies with 'rat-tailed maggots') are relatively catholic in their habitat preferences, but some such as *Eristalis cryptarum* and *E. rupium* are more specialised.

The adults of many wetland hoverflies, such as *Helophilus pendulus*, *Eristalis pertinax* and *E. tenax*, occur well away from breeding sites, although their ubiquitous occurrence may also suggest that they can use small patches of habitat that we overlook. This behaviour means that it is not always necessary to visit what are often thought to be the ideal places for hoverflies. Roadside verges often prove to be very productive, especially if there are lots of nectar sources. Indeed, in areas where there are large expanses of intensively managed agricultural land or heavily grazed moorland, roadside verges may provide the only suitable habitat.

Conifer plantations can also be very productive, especially in the uplands, where it is not uncommon to find rich assemblages of hoverflies. The reasons for this lie in the way that conifer plantations have changed the upland environment. Former sheepwalks often revert to heathland within woodland

A ditch on a coastal marsh on which *Lejops vittatus* (MIDDLE) occurs.

rides, and where there are wide margins, the resulting vegetation can be extremely floriferous, sheltered and protected from sheep grazing. Consequently, conifer plantations have plenty of lures to attract hoverflies and the hoverfly recorder.

Finally, it is worth remembering that female hoverflies need to eat pollen to provide protein for egg production and will often visit pollen-rich sources such as grasses and plantains. A list of hoverflies for any site can be greatly enhanced by sweeping a verge with flowering Ribwort Plantain. Apart from the common *Melanostoma* species, numbers of *Platycheirus* will occur, especially *P. clypeatus*, *P. angustatus* and *P. manicatus*. Sweeping is also a very effective way of finding *Sphegina* in northern and western Britain, and in Ireland, especially where Pignut abounds: occasionally three, or even all four, species in this genus can be found at the same locality, whereas visual searches may be far less successful.

Finding *Parasyrphus nigritarsis*

Parasyrphus nigritarsis (p. 144) was considered a great rarity until its larvae were found to feed on the larvae of the Green Dock Beetle *Gastrophysa viridula*. The beetle lays a clutch of bright orange eggs on the underside of dock leaves. The hoverfly's white eggs can be seen nestling among them. Later, the hoverfly larvae can be found with the beetle larvae by examining the underside of dock leaves (note that the beetle larvae create obvious holes in the dock leaves).

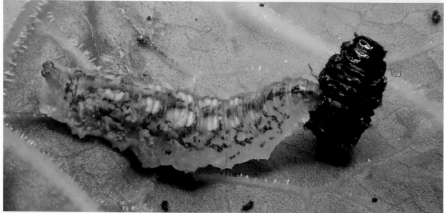

A *Parasyrphus nigritarsis* egg among a mass of eggs of the Green Dock Beetle (TOP) and a larva consuming a beetle larva (BOTTOM).

The search for *Callicera rufa*

Adults of *Callicera* are very elusive. It is thought that they spend much of their time high up in the canopy and that the only times they are to be found at ground level are when they emerge, if they need to come down to drink or when females lay eggs. Consequently, the spectacular *C. rufa*, a resident of the Caledonian pine forests of central Scotland, was recorded on very few occasions before 1988 and was believed to be a great rarity.

Callicera rufa

In the late 1980s work by the Malloch Society showed that *C. rufa* was much easier to find by looking for larvae in water-filled rot-holes in Scots Pine. During the course of a few field seasons they were able to record it from almost all sites in the Scottish Highlands where Caledonian pine forest still existed. It was also found in suitable rot-holes in mature Scots Pine plantations and even occasionally in spruce and larch. More recently, larvae have been found in water-filled cavities in the stumps remaining after mature plantations have been felled.

Artificial cavities in stumps, created using a chainsaw, will naturally fill with rainwater. They are often occupied the same year – making artificial rot-holes a great way to find new populations. This has been proven in England since *C. rufa* was first discovered in Sherwood Forest in 2009. The origins of this range expansion are unclear: was it southward movement from Scotland or colonisation from Europe?

Artificial rot-hole created in the stump of a Scots Pine using a chainsaw.

Before 1988

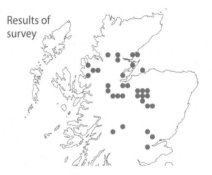

Results of survey

Glossary

Diagrams showing the locations of terms in **bold brown text** are on *pages 48–52*.

adpressed	pressed closely against, or lying flat
antenna (plural: **antennae**)	the 'feelers' on the front of the head of an insect. These bear chemo-sensory organs and allow the detection of chemical stimuli such as the odours emitted by sap or fungi – see *page 49*
aphidophagous	feeding on aphids
apical	in the direction of the apex. Hence, that part of an appendage lying nearest its tip and farthest from the point of attachment – also **distal** (opposite of **basal**)
aquatic	living in water. Strictly, this covers any type of water, fresh or salt, but when applied to Diptera it usually means fresh water
arista (plural: **aristae**)	appears as a bristle arising from the surface of the 3rd antennal segment. It is actually the remnant of antennal segments and can show signs of segmentation. It can be bare, clothed in short hairs ('**pubescent**') or long hairs ('**plumose**') – see *page 49*
basal	in the direction of the base. That part of an appendage nearest to the point of attachment – also **proximal** (opposite of **apical**)
cell	an area of the wing membrane bounded by the wing margin and/or veins. Wing cells are named after the vein that precedes them (*i.e.* in front) and are given lower-case abbreviations (*e.g.* 'cup')
chitin	the tough, protective, semi-transparent substance that forms a hoverfly's body, wing veins, *etc.*
compound eye	eye made up of large numbers of **ommatidia**. In the Diptera, the compound eyes normally occupy a large part of the head
costa	the main vein forming the leading edge of the wing
coxa (plural: **coxae**)	the basal segment of a leg – the part attached to the thorax
cross-vein	short veins connecting the length-wise veins and their branches
cuticle	the hard, protective layer that forms the outer surface ('skin') of an invertebrate
dimorphic	occurring in two distinct forms
dimorphism	a difference in size, form or colour between individuals of the same species, characterising two distinct types
discal cell	a closed cell in the centre of the wing bordered by **M** veins and closed by cross-vein **R–M** – see *page 50*
distal	farthest from the mid-line of the body. Another term for **apical**
dorsal	on the upper surface
dorsum	the dorsal surface, usually of the **thorax** – hence **thoracic dorsum**
dusting	a characteristic of the surface of the chitinous plates making up the body of a hoverfly. 'Dust' is actually formed by minute, flattened hairs rather like the scales of Lepidoptera (butterflies and moths) – see *page 52*
entomophagous	growing in or on an insect, for example certain fungi
face	the plate that forms the front of the head, delineated by the antennal sockets above, the mouth opening below and, laterally, by the compound eyes
femur	the principal leg segment, analogous to the 'thigh', located between the **trochanter** and the **tibia**
frass	the droppings of plant-eating (phytophagous) insects
frons	the plate forming the top of the head, bordered by the compound eyes, the **ocellar triangle** and the antennal bases

GLOSSARY

genitalia	the copulatory organs. The shape and arrangement of the genitalia are often used to distinguish between closely related or very similar species
haltere	remains of the hind wing of Diptera which has become an organ of balance
humerus (plural: **humeri**)	the anterior corners or 'shoulders' of the **thoracic dorsum** – see *page 48*
imago	adult
instar	the stage in an insect's life history between any two moults. A newly hatched insect that has not yet moulted is said to be a first-instar larva. The adult (**imago**) is the final instar
integument	the 'skin' or outer membrane
jizz	the often indefinable characteristic impression given by an animal or plant, usually defined by shape or movement
larva (plural: **larvae**)	the immature form of an insect which is markedly different from other life-stages such as the **pupa** or adult (**imago**)
lunule	a small area on the frons just above the antennae
malaise trap	a large, tent-like structure used for trapping flying insects, particularly Diptera and Hymenoptera
metatarsus	the most basal of the five **tarsal segments**; also called the 'basitarsus' – see *page 51*
microtrichia	the microscopic hairs on the surface of the wing membrane
occiput	the back of the head, behind the compound eyes
ocellar triangle	the plate on which the ocelli are located. Usually a sharply delineated, triangular area of the **frons** – see *pages 48, 49, 52*
ocellus (plural: **ocelli**)	simple eye. Nearly always three, arranged in a triangle at the **vertex** of the head and located on a sharply delineated triangular plate – the **ocellar triangle**
ommatidium (plural: **ommatidia**)	one of the individual structural elements of the **compound eye** of an insect
oviposition	the act of laying eggs
ovipositor	the egg-laying structure of the female
petiole	a slender stalk between two structures. In this context: the stalk formed where wing veins join, which then runs to the wing margin (see, for example, the illustration on *page 237*)
phytophagous	feeding on plants
pilosity	hairiness
pleura (singular: **pleuron**)	the plates making up the sides of the **thorax**
plumose	with long hairs or bristles. Usually applied to the **arista**
polymorphic	occurring in several distinct forms – see *page 28*
porrect	extending horizontally, not drooping
posterior	at the back end, towards the tail; the rearward-facing surface of a structure
proboscis	in Diptera this term refers to the extensile (extendable) mouthparts
proximal	another term for **basal**
pubescent	with short hairs. Often applied to the **arista**
pupa	the stage in a hoverfly's life-cycle, often quiescent (inactive), that precedes the emergence of the adult (**imago**)
puparium	the pupal **integument** or shell

GLOSSARY

rostrum	a snout-like projection of the head
saprophagous	feeding on dead and decaying organic matter
scutellum	the shield-shaped, posterior part of the **thoracic dorsum**
scutellar hairs and bristles	the marginal hairs and bristles near the apex of the **scutellum**
spiracle	an air inlet for the insect's breathing system (in the Diptera, the **thorax** has two spiracles on each side)
spurious vein	See **vena spuria**
squama (plural: **squamae**)	two (upper and lower) lobes at the base of the hind margin of the wing adjacent to the **halteres**
sternite	the plate forming the bottom, or ventral surface, of a segment of an insect's body. Usually refers to the underside of an abdominal segment when used in keys to Diptera
stigma	coloured area of the wing next to the **costa** and near the end of veins **Sc** or **R₁** – see *page 50*
subcosta (Sc)	the second, usually unbranched, longitudinal wing vein, posterior to the **costa**
sweeping	sweeping a net gently back and forth through low vegetation
synanthropic	mainly occurring in habitats created by humans
tarsus	the apical (outermost) part of a leg, consisting of five segments (tarsal segments or 'tarsomeres'). The first segment is termed the 'metatarsus' or 'basitarsus' and the fifth carries the claws – see *page 51*
taxon (plural: **taxa**)	a general term for a unit of biological classification. Often used as an equivalent to species, but this is not really accurate since it can refer to any level in the taxonomic hierarchy, *e.g.* a genus or a subspecies
tergite	the plate forming the top, or **dorsal** surface, of a segment of an insect's body. Usually refers to the top of an abdominal segment when used in keys to Diptera. These segments are referred to as T1 to T5 in the text
terminal	at the end
thoracic dorsum	the **dorsal** surface of the thorax
thorax	the central division of the body of an adult insect, consisting of three fused segments each of which bears a pair of legs and the hind two of which bear a pair of wings, when present
tibia	the fourth segment of a leg, analogous to the 'shin', between the **femur** and the **tarsus**
trochanter	the small, second segment of a leg between the **coxa** and the **femur**
vagrant	an individual that wanders outside the normal range of its species
vein	chitinous, rod-like or hollow tube-like structure supporting and stiffening the wings in insects, especially those extending longitudinally from the base of the wing to the outer margin
vena spuria	a longitudinal fold in the wing in the family Syrphidae which is chitinised along its crease. It runs between the **R** and **M** veins and crosses **R–M**. Although it looks like a vein it is not connected to the rest of the **venation**
venation	the arrangement of the wing veins – see *page 50*
ventral	on the lower surface
vertex	the highest point (especially of the head); the apex
zygoma	a sharply defined region of the face running along the eye margin; a feature of the tribe Cheilosiini – see *pages 59, 170*

Identifying hoverflies

Identification skills take time to develop. New recorders typically start with conspicuous species and encounter a wider range of species as their interest develops. A few hoverflies can be identified readily by sight in the field, even from some distance, whilst more species can be identified in the field with a little more practice and experience: that is what this book is primarily about. As your interest grows you will need to develop new field skills, as many hoverflies are not likely to be found without a targeted search.

♀ *Ferdinandea cuprea* – one of the more straightforward hoverflies to identify.

Having found your hoverfly, you need to get close (at least 50 cm – a foot or two) in order to see the necessary features. For example, among the common, bee-like *Eristalis* you need to find out whether it has a black stripe down the middle of the face, what colour its front tarsi are, and so on. Providing you can get a good, close look as it moves about you may be able to see the necessary features. Close-focusing binoculars can help and many modern roof-prism binoculars are capable of focusing down to 1 m or less. Where even more detailed examination is necessary, it is best is to catch the hoverfly in a net, pick it out carefully between your thumb and forefinger and examine it closely, with the aid of a 10× hand lens (see *page 317*). It may be helpful for photographers to try to get images of the undersides of some tricky species (*e.g.* male *Parhelophilus* and *Eumerus*, most *Neoascia*, female *Paragus* and the *Dasysyrphus venustus* complex) in the hope of depicting critical features that are not seen from above. Using the sorts of tips and tricks covered by this book, with sufficient experience you should be able to identify at least 100 or so of the British and Irish species in this way. Once you have checked the fly over, it can be released unharmed.

At least half of the British and Irish hoverfly species cannot be identified without taking a specimen home for examination under good lighting using a binocular microscope at around 10–30× magnification. In some cases, good photographs circumvent this need but there are many that cannot be recognised from even the best photos. The identification of these species goes beyond the scope of this book. *British Hoverflies* (second edition) by Stubbs and Falk (2002) is the best available UK text for this purpose but Bot & Van de Meutter (2023) and van Veen (2004) are very useful, too – see *Further reading* (*page 337*).

Identification also requires an understanding of a whole new terminology, describing the insect's body and its features. This book tries to explain and illustrate such terms, but there are many tips and tricks involving the handling, positioning and lighting of specimens that it is much easier to demonstrate than to describe in words – which is where training courses can help. Local Records Centres and the Field Studies Council are potential providers of such training.

IDENTIFYING HOVERFLIES

Naming the parts

Knowledge of the terminology used when describing the anatomy of a hoverfly is important in understanding the descriptions given in the species accounts in this book.

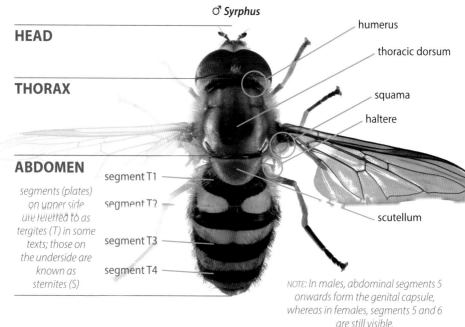

♂ *Syrphus*

HEAD

THORAX

ABDOMEN

segments (plates) on upper side are referred to as tergites (T) in some texts; those on the underside are known as sternites (S)

- segment T1
- segment T2
- segment T3
- segment T4

- humerus
- thoracic dorsum
- squama
- haltere
- scutellum

NOTE: *In males, abdominal segments 5 onwards form the genital capsule, whereas in females, segments 5 and 6 are still visible.*

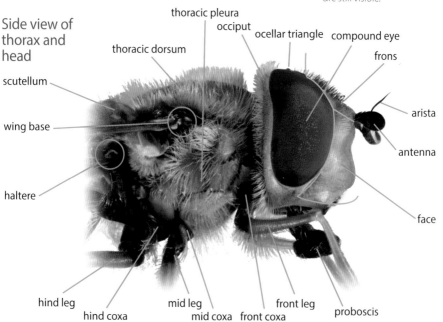

Side view of thorax and head

- scutellum
- wing base
- haltere
- thoracic dorsum
- thoracic pleura
- occiput
- ocellar triangle
- compound eye
- frons
- arista
- antenna
- face
- hind leg
- hind coxa
- mid leg
- mid coxa
- front leg
- front coxa
- proboscis

IDENTIFYING HOVERFLIES

HEAD

The head is dominated by large compound eyes. In males these are typically touching at the top of the head, but in females they are separated. If the eyes are touching, then you can be sure it is a male, but if they are separated the sex of the individual depends on the genus as there are a few genera for which this rule does not apply. (e.g. *Anasimyia*, *Helophilus*, *Lejogaster*, *Lejops*, *Microdon*, *Neoascia*, *Parhelophilus*, *Pelecocera* and *Sphegina*). The area on top of the head between the eyes is the **frons** (which is rather small in the males of those species where the eyes touch). Right at the top of the head there are three simple eyes, or **ocelli**, in a triangular formation, usually on a slightly raised area known as the **ocellar triangle**. The ocelli are not capable of image resolution, but they are sensitive to light and are used to measure day length and regulate a hoverfly's internal clock.

A pair of **antennae** are situated below the front end of the frons. Each consists of three segments (conventionally numbered from the base outwards), with the third segment usually being the largest. It varies considerably between species, and the shape and size of the antennae is often used in descriptions. The third segment bears the **arista**, which usually arises from the top surface somewhere between the base and the middle, when it is described as dorsal. However, in some the arista arises from the tip, in which case it is described as apical. The aristae may be bare or hairy: if it has very long hairs so that it looks like a feather, or a TV aerial then it is described as **plumose;** if the hairs are short then it is **pubescent**; if hairs are absent then it is referred to as **bare**.

The **face** is below the antennae and between the eyes. It occupies the area between the base of the antennae and the mouth margin. Its colour is often useful for identification. The face often has a nose-like bulge or 'knob' in the middle and the presence or absence and shape of this is frequently mentioned in descriptions; to appreciate this feature you need to view the hoverfly's head in profile. The frons or face may be dusted – see *page 52*.

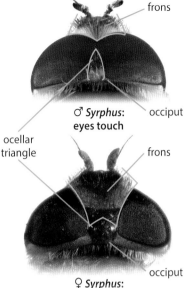

♂ *Syrphus*: eyes touch

♀ *Syrphus*: eyes separated

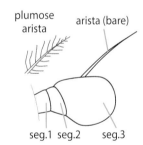

'Normal' antenna segments

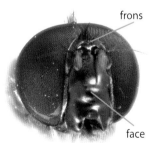

Head of *Cheilosia*

THORAX

The thorax bears the wings and three pairs of legs and also, just behind and below the wing bases, the **halteres**. The top of the thorax is the **thoracic dorsum**, the colour and pattern of which is sometimes helpful in identification. At each front corner there is a swelling at the shoulders; together these are called the **humeri**. Behind the dorsum is a semicircular swelling termed the **scutellum**. The sides of the thorax are termed the **thoracic pleura**.

IDENTIFYING HOVERFLIES

Wings

The wings consist of a transparent membrane supported by a series of struts: the **veins**. The main veins run from the base of the wing towards the tip and are occasionally linked by **cross-veins**.

The naming of wing veins in flies is a complicated subject and many systems have been devised. Unfortunately, these various systems have sometimes used the same name for different veins, which can be very confusing when comparing descriptions in different books! The system adopted here (the Comstock-Needham system) is also used in *British Hoverflies* by Stubbs and Falk (2002) and most of the more recent European works. It recognises the following main veins running from the wing base (from front to back): the **costa** (C), **sub-costa** (Sc), **radial vein** (R), **medial vein** (M) and **anal vein** (A). It is thought that, in the most primitive flies, these veins had a series of branches, but that some of these branches have fused together again during the course of later evolution. As a result, seemingly strange labels such as R_{2+3}, R_{4+5} and M_{1+2} are used to refer to veins that are believed to derive from these fusions. Where a section of the wing membrane is surrounded by veins (or by veins on three sides and the margin of the wing on the other) it is called a **cell**. Cells are also given names.

The most important feature to recognise is the **R–M cross-vein** in the middle of the wing. This is always present in hoverflies and is almost always near to or at the middle of the wing. R–M arises from vein R_{4+5} and forms the outer border of the **1st basal cell**. The R–M cross-vein, together with the **2nd basal cell** (the one immediately below the 1st basal cell and with three veins arising from its outer end) and the cell below it (the **discal cell**) are the main features you need to be able to find (see *page 54*) in order to follow the descriptions.

There is often a coloured area of wing membrane between the tip of the sub-costa and the tip of vein R_1, termed the **stigma**. Some species also have a strong darkening across the middle of the wing membrane, referred to as a **wing cloud**. This shading often extends from just behind the stigma, around the R–M cross-vein and over the end of the discal cell.

The wing membrane is usually covered in tiny hairs called **microtrichia**. To see these, high magnification (around 30–40×) is needed, with the light coming through the wing membrane from behind. In many cases, microtrichia do not cover the entire wing surface, and the patterns they make are useful characters in the identification of some difficult species. The degree to which the 2nd basal cell is covered is most commonly used.

'Normal' hoverfly wing venation (*Syrphus*)

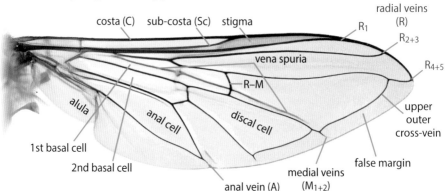

IDENTIFYING HOVERFLIES

Legs

Each leg is attached to the thorax by a **coxa** and a small segment termed the **trochanter**.

The two main parts of the leg, the **femur** and the **tibia**, have the same names as the two main leg bones of a human and, like our legs, have a 'knee' joint between them. Finally, the hoverfly's equivalent of our foot is composed of five **tarsal segments** or **tarsi**. The first of these is usually the longest and is called the **metatarsus**. The last tarsal segment (seg. 5) bears a pair of claws.

When species descriptions refer to the **base** of a leg joint, they mean the part nearest to the body, so the image (left) could be described: hind femur mainly yellow, black only at extreme base. The opposite is the **apex**, or '**apical**', which is the part farthest from the body.

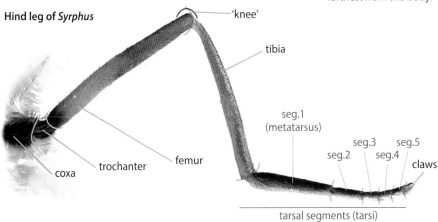

Hind leg of *Syrphus*

ABDOMEN

The abdomen is composed of a series of segments (or **tergites** (T)), which are numbered from the base (where the abdomen joins the thorax) towards the tip. The 1st segment (T1) is often inconspicuous, being largely hidden by the **scutellum**. Consequently, the 2nd segment (T2) is usually the first that is obvious when viewed from above. There are coloured markings on some or all the tergites in certain species, the shape, colour and positions of which are often used in identification. The male **genitalia**, positioned at the tip of the abdomen, are often obvious as a capsule, folded under the end of the abdomen and forming a distinct bulge in side view. The abdomen of the female usually tapers to a blunt, conical point, with no trace of a bulge, providing another way of determining the sex of a hoverfly in species where the males' eyes are not touching. Very occasionally the colour or dusting of the plates on the underside of the abdomen is used in species descriptions. These plates are termed **sternites** (S).

Abdomen of ♂ *Syrphus*

genital capsule (♂ only)

IDENTIFYING HOVERFLIES

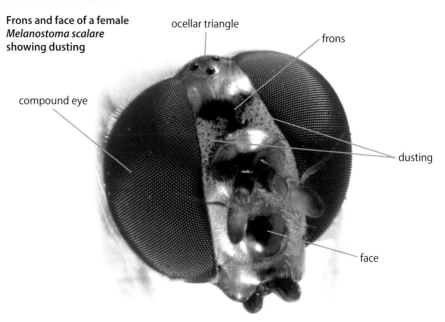

Frons and face of a female *Melanostoma scalare* showing dusting

ocellar triangle

frons

compound eye

dusting

face

Dusting

Patterns and markings in hoverflies are of two kinds: coloured areas of the **cuticle** and patterns formed on the surface by bands of differently coloured hairs and **dusting**.

Dusting is actually formed by tiny flattened hairs. The name is quite descriptive: it looks like patches of dust on the shiny surface of the hoverfly's cuticle. The characteristic feature of dusting is that its appearance depends on the lighting. As you move a hoverfly around so that the light comes from different directions, the appearance of dusting changes. At some angles it may almost disappear, whilst at others it may be obvious, contrasting with the shiny cuticle.

The appearance of patterns formed by bands of coloured hairs changes in a similar way according to the direction and intensity of the light. It can often be difficult to assess the colour of hairs, and fine black hairs can appear to be pale when brightly lit. This is because you are actually seeing the bright reflections off the surface of individual hairs rather than their true colour.

By contrast, markings due to coloured patches of the insect's cuticle do not tend to change in appearance as the direction of lighting changes.

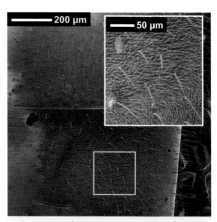

Scanning electron micrographs of dust patches on the abdomen of a female *Platycheirus albimanus* showing that they are composed of tiny hairs.

Guide to the tribes

Hoverflies (family Syrphidae) are divided into 'tribes' by Stubbs and Falk (2002) in order to make keys more accessible. Although the concept of tribes is not used in readily available European literature, for the purpose of this book, it is helpful from an organisational perspective to split the family into a series of subfamilies, tribes and then genera. **A summary of the features of British and Irish hoverfly tribes, arranged under the three subfamilies (Syrphinae, Eristalinae and Microdontinae) appears on** *page 61*.

This section provides a guide to the characters that are best used as pointers to identify tribes, and in some cases genera or species within that tribe. It focuses on features that can be seen in the field or by using a 10× hand lens. Experience will undoubtedly help in speeding up this process, but faced with an unfamiliar hoverfly, the following is a suggested process to assist identification.

i)	Confirm that it is a hoverfly by the presence of the **vena spuria** (see *p. 50*) – [except for *Psilota* – see **2a** (*p. 54*)]	
ii)	Establish whether the front of the thorax behind the head is visible or obscured and whether the humeri are hairy or bare – see **1** *below*	
iii)	**1a If they are visible and hairy:** look initially at the wing venation and, from there, additional characteristics of the face, antennae and aristae. – see **2** – **7** (*pp. 54–59*)	**1b If they are obscured and/or bare:** look at additional features of the **Syrphini**, **Bacchini** and **Paragini**. – see **8** – **9** (*p. 60*)
iv)	Once the tribe has been established using the keys go to the relevant *Guide to the tribe* page given to identify the genus.	

1

a) **Front of the thorax generally visible, making it possible to see the humeri – which are hairy**

Very variable in form, and includes the majority of the big bee and wasp mimics.

Most hoverfly tribes. ▶ **2**

b) **Head concave, making it hard to see the front of the thorax and obscuring the humeri – which are bare**

Includes the majority of black-and-yellow hoverflies; most (except **Bacchini**) with at least some yellow on their face.

Syrphini	Bacchini	Paragini	▶ **8**
p. 98	*p. 72*	*p. 96*	

ID GUIDE TO THE TRIBES

2
from 1a

a) Vena spuria absent

The presence of a vena spuria is a feature of all British and Irish hoverflies, except for one species, and can be fundamental to the identification of those species that mimic or look very similar to Hymenoptera or other Diptera.

The species that lacks a vena spuria is *Psilota anthracina* (but this is absent from Ireland). Beware – great care should be taken with its identification as this rare species is easily confused with some blue-black muscid flies.

▶ **Merodontini:** *Psilota* p. 250

b) Strong loop in wing vein R_{4+5}

A group of mainly medium-sized hoverflies including convincing honey-bee and bumblebee mimics.

Beware – some **Syrphini** e.g. *Didea*, the subgenus *Lapposyrphus* (within **Eupeodes**) and *Megasyrphus* have a dip in R_{4+5}. See *page 99*. However, checking the head and humeri (see **1b**) should avoid any confusion.

▶ **Eristalini** *p. 218*
▶ **Merodontini:** *Merodon p. 248*

c) 'Normal' wing – no wing loop present

▶

'Normal' hoverfly wing venation (*Syrphus*) (from *page 50*); features in the key are in **bold text**.

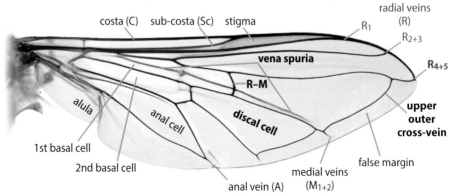

GUIDE TO THE TRIBES

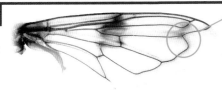

a) **Upper outer cross-vein re-entrant**
(turning back towards the body)
This is a small group of hoverflies which includes some of the largest bee and wasp mimics.

▶ **4**

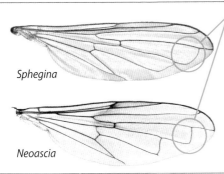

Sphegina

Neoascia

b) **Upper outer cross-vein strongly upturned**
(but not turning back towards the body)
Small, narrow-bodied hoverflies
This feature is shared by two very similar genera – *Sphegina* and *Neoascia*.

▶ **Chrysogastrini:**
Sphegina p. 198
Neoascia p. 200

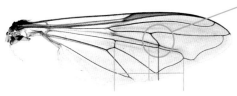

c) **'Normal' wing – upper outer cross-vein neither re-entrant nor strongly upturned;**
Inner cross-vein R–M meets the discal cell before the middle of the cell

▶ **5**

d) **As c) but the inner cross-vein R–M meets the discal cell at a point at or beyond the middle of the cell**
A heterogeneous group of bumblebee and honey-bee mimics, together with more elongate species which resemble some sawflies.

▶ **Xylotini** *p. 276*

Ferdinandea cuprea

Xylota sylvarum – a sawfly mimic

Beware – *Ferdinandea* [**Cheilosiini**] (p. 192) has similar wing venation, but looks very different, with a brassy, metallic abdomen; grey stripes running along the thorax; and a few bristles on the upper sides of the thorax.

ID GUIDE TO THE TRIBES

4 from 3a

a) Aristae strongly plumose

Large bumblebee and wasp mimics.

▶ **Volucellini:** *Volucella* p. 270

b) Antennae long and forward-pointing ('porrect')

Medium-sized dumpy hoverflies with relatively short wings.

▶ **Microdontinae:** *Microdon* p. 298

c) Antennae 'normal' with a bare arista
Hind femur enlarged

Small, shiny or brassy hoverflies.

▶ **Merodontini:** *Eumerus* p. 246

---

4 Hoverflies with wings which have the upper outer cross-vein re-entrant

Microdontinae: *Microdon analis* p. 298   **Merodontini:** *Eumerus* sp. p. 246

Bumblebee mimic
Volucellini: *Volucella bombylans* p. 270

Wasp mimic
Volucellini: *Volucella inanis* p. 274

See also *Identifying wasp and bee mimics* pp. 62–65

GUIDE TO THE TRIBES

5 *from 3c*

a) **Antennae 'porrect' with a terminal arista (which is white-tipped)**

▶ **Callicerini:** *Callicera* p. 164

The three species of *Callicera* are all rare and might be overlooked as solitary bees.

Check – wing venation (presence or absence of a vena spuria) to ascertain whether it is a hoverfly or solitary bee.

arista antenna

b) **Aristae not terminal; strongly plumose Large hoverflies**

▶ **Sericomyiini** p. 266

▶ **Bumblebee mimic** *Sericomyia suberbiens* p. 266
▶ **Wasp mimics** *S. lappona* & *S. silentis* p. 268

Beware – *Volucella* (p. 270) are also large with plumose aristae but with different wing venation (**3a**) and should already have keyed out at **4a**.

arista antenna

c) **Aristae not terminal; bare or with weak pilosity Antennae various shapes Generally much smaller species**

▶ **6**

arista antenna

5b Hoverflies with strongly plumose aristae and 'normal' wings

Bumblebee mimic
Sericomyiini: *Sericomyia superbiens* p. 266

Wasp mimic
Sericomyiini: *Sericomyia silentis* p. 268

See also *Identifying wasp and bee mimics* pp. 62–65

ID GUIDE TO THE TRIBES

6
from 5c

a) Face flat, with long drooping hairs

This is a group of mainly black hoverflies, some with yellow spots on abdomen segment T2.

▶ **Pipizini** *p. 254*

Beware – *Psilota anthracina* (*p. 250*) may also key out here (see **2a**). It is a shining bluish-black hoverfly whose flat face has a strongly pointed mouth-edge.

face

b) Face strongly projected

▶ **Cheilosiini:** *Rhingia* p. 194

Note: *Anasimyia lineata* [*Lristalis*] (*p. 240*) has a similar but less extreme projection, although the strong wing loop distinguishes this species.

face

c) Face with some projection; Antennae unusual, half-moon-shaped and with a strongly thickened arista

▶ **Pelecocerini:** *Pelecocera* p. 252

These are relatively small black-and-yellow hoverflies that are confined to the heathlands of southern England and to Scottish conifer woods.

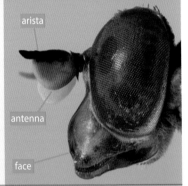
arista
antenna
face

d) Face with some or no projection; Antennae 'normal' with a fine arista

▶ **7**

Note: The species illustrated is *Portevinia maculata* – note the zygoma which is diagnostic of the Cheilosiini (see **7a**).

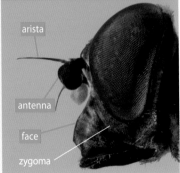
arista
antenna
face
zygoma

GUIDE TO THE TRIBES

7a) Cheilosiini: *Cheilosia bergenstammi*   **7b) Chrysogastrini:** *Lejogaster metallina*

7
from 6d

a) Face with a 'nose' above the mouth – giving the face a somewhat bulbous appearance; zygoma present

▶ **Cheilosiini** *p. 168*

A heterogeneous group of hoverflies: most *Cheilosia* are black and unmarked, although some are very hairy, and are weak bee mimics; *Portevinia* (p. 192) has distinct markings; both *Rhingia* (see **6b**) and *Ferdinandea* (p. 192) are colourful and distinctive.

b) Face with no 'nose' – giving the face a strongly concave appearance; zygoma absent

Beware – some species have a slight 'nose', but never as distinct as in **Cheilosiini** (p. 168).

▶ **Chrysogastrini** [part] *p. 196*

A heterogeneous group of dark hoverflies, many of which have a metallic sheen under certain light conditions.

c) Face without projection or bristles:
Wholly brownish-orange
– can darken with age

▶ **Chrysogastrini:** *Hammerschmidtia p. 216*

With a grey thorax and brownish-orange abdomen

▶ **Chrysogastrini:** *Brachyopa p. 212*

Beware – many species in other fly families are similar in appearance – check the wing for a vena spuria to make sure it is a hoverfly!

Brachyopa scutellaris

ID GUIDE TO THE TRIBES

Hoverflies with head concave, making it hard to see the front of the thorax and obscuring the humeri – which are bare

8 from **1b**

a) Face ground colour black (although there may be paler dusting over the black ground colour)
   Scutellum black
   ▶ **Bacchini** *p. 72*

Melanostoma mellinum

b) Face partially or wholly yellow
   Scutellum either black or yellow
   ▶ **9**

Paragus [**Paragini**] (*p. 96*) can have quite a dark face with only weak yellow areas towards the edges.

9 from **8b**

a) Tiny black flies with a distinctly narrowed 'waist' to the abdomen
   Body length less than 5 mm
   Face yellow with central black stripe.
   ▶ **Paragini:** *Paragus* p. 96

b) Mainly colourful flies
   Body length 5–12 mm
   Both face and scutellum usually at least partially yellow, but a small number of species are darker *e.g. Leucozona laternaria* (*p. 118*) [black scutellum]. Some *Melangyna* (*pp. 154–159*) [very dark faces].
   ▶ **Syrphini** *p. 98*

Simplified guide to British and Irish hoverfly tribes

Showing the set of key features to look for.

The figures in the Gen/Spp. column indicate the number of genera within each tribe, the number of species illustrated in the book, and the [total number of species within the tribe].

SYRPHINAE:	THORAX: front obscured; humeri bare:	Gen/Spp.
BACCHINI p.72	FORM: generally somewhat elongate FACE black SCUTELLUM: black	4 genera 15 [30] spp.
PARAGINI p.96	FORM: tiny (≤5mm) black FACE yellow, often obscure, but always with central black stripe	1 genus 1 [4] spp.
SYRPHINI p.98	FORM: larger (≥5–12mm); colourful FACE some yellow, some with a central black stripe, some black SCUTELLUM: often yellow, at least in part	18 genera 58 [85] spp.
ERISTALINAE:	**THORAX: front visible; humeri hairy**	**Gen/Spp.**
ERISTALINI p.218	WING strong wing loop R$_{4+5}$ (cf. Merodontini: Merodon) LEGS: black/yellow	8 genera 22 [?8] spp.
VOLUCELLINI p.270	WING outer upper cross-vein re-entrant ARISTA: plumose (cf. Merodontini: Eumerus; Microdontinae)	1 genus 5 [5] spp.

MERODONTINI p.246	Heterogeneous tribe: ANTENNA & ARISTA: 'normal'
	   Psilota – WING vena spuria absent
	   Merodon – WING strong wing loop R$_{4+5}$ (cf. Eristalini)
	                LEGS: all black
	   Eumerus – WING outer upper cross-vein re-entrant
	                LEGS [hind femur]: enlarged
		3 genera
		6 [7] spp.

XYLOTINI p.276	wing 'normal' — + the inner cross-vein R–M meets the discal cell at a point at or beyond middle of the cell	10 genera 20 [21] spp.
CALLICERINI p.164	defined as one with no wing loop in R$_{4+5}$ and the upper outer cross-vein neither re-entrant nor strongly upturned – see page 50 and **3c**, page 55 — ANTENNA: porrect / ARISTA: terminal with white tip	1 genus 3 [3] spp.
SERICOMYIINI p.266	WING outer upper cross-vein not re-entrant / ARISTA: plumose	2 genera 3 [3] spp.
PELECOCERINI p.252	ANTENNA: half-moon-shaped / ARISTA: thickened	1 genus 2 [3] spp.
PIPIZINI p.254	FACE flat with long drooping hairs / FORM: small–medium, predominantly **black**	5 genera 10 [20] spp.
CHEILOSIINI p.168	FACE nose-like central prominence; zygoma present	4 genera 21 [43] spp.
CHRYSOGASTRINI p.196	Heterogeneous tribe with no consistent features; predominantly small–medium-sized dark hoverflies with concave faces; some spp. metallic; a few spp. colourful – see *Guide to Chrysogastrini* for more information (p. 196)	10 genera 19 [29] spp.
MICRODONTINAE: p.298	WING outer upper cross-vein re-entrant ANTENNA: porrect (cf. Volucellini, Merodontini: Eumerus)	1 genus 4 [4] spp.

Identifying wasp and bee mimics (see also *page 30*)

The following pages cover the identification of hoverflies that mimic members of the Hymenoptera (such as bees, bumblebees and wasps), although the quality of the mimicry is sometimes open to question! If there is any doubt as to whether an individual is a hoverfly or a hymenopteran, look at the shape and details of the antennae. Typical examples are illustrated here.

Hoverfly (Diptera) — 3 antennal segments

Hymenopteran — long, filamentous

Wasp mimics

WING VENATION see *pp. 54–55*

Social wasp/hornet mimics

ARISTA **plumose**

WING upper outer cross-vein re-entrant; HIND LEGS dark Beware – care is needed to separate the two *Volucella* spp.		WING 'normal', upper outer cross-vein not re-entrant; HIND LEGS brown
ABDOMEN UNDERSIDE S2 black	ABDOMEN UNDERSIDE S2 yellow	

Volucella zonaria p. 274

Volucella inanis p. 274

Sericomyia silentis p. 268

Social wasp mimics
ARISTA **porrect**

Chrysotoxum spp. (especially *C. cautum*) pp. 31, 102–107

NOTE: *Chrysotoxum* spp. make a very good job of mimicking a social wasp's 'kneed' antennae in side view, especially when the 3rd segment is held drooped at an angle to the first two. The rather arched shape of the abdomen also adds to the deception.

Wasp mimics
ARISTA **'normal'**

most other black-and-yellow hoverflies

NOTE: Many species have yellow bands or spots and appear to mimic social and solitary wasps. The effectiveness of this mimicry is debatable, but it must be considered from the perspective of an animal in flight where the movement forms part of the illusion.

Wasp species that hoverflies mimic: Hornet *Vespa crabro* (TOP) and Common Wasp *Vespula vulgaris* (BOTTOM)

Honey-bee and other bee mimics

IDENTIFYING WASP AND BEE MIMICS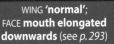
WING VENATION see pp. 54–55

Honey-bee mimics

WING **loop in vein R4+5 present**		WING **'normal'**;
LEGS **hind tibia curved and enlarged, with stiff hairs that resemble a pollen basket**	LEGS **hind femur enlarged, hind tibia not curved**; FORM **greenish hue**	FACE **mouth elongated downwards** (see p. 293); LEGS **hind femur not enlarged**

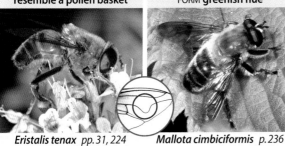

Eristalis tenax pp. 31, 224 — *Mallota cimbiciformis* p. 236 — *Criorhina asilica* p. 292

NOTE: Other *Eristalis* (see p. 220) are less convincing bee mimics.

NOTE: All *Criorhina* (see p. 292) are convincing bee mimics.

Mining bee mimic

WING **'normal'**; FACE **zygoma present** – diagnostic of genus (see p. 170)

Solitary bee/honey-bee mimics

WING **'normal'**; LEGS **hind femora enlarged**

LEGS **hind femora enlarged (arched in ♂)**; THORAX **underside bare**	LEGS **hind femora enlarged (straight in both sexes)**; THORAX **underside hairy**

Cheilosia chrysocoma p. 186 — *Brachypalpus laphriformis* p. 282 — *Chalcosyrphus eunotus* p. 284

NOTE: Other furry *Cheilosia* – *C. albipila* (p. 184) and *C. grossa* (p. 184) – are sometimes described as bee mimics but are unconvincing.

NOTE: See p. 283 for a comparison of thorax undersides.

Three bees that hoverflies mimic: Honey-bee *Apis mellifera* (LEFT), a mason bee *Andrena carantonica* (CENTRE) and a mining bee *Osmia bicornis* (RIGHT)

63

IDENTIFYING WASP AND BEE MIMICS

Bumblebee mimics

WING VENATION see pp. 54–55

Once established as a hoverfly, the bumblebee mimics can present an identification challenge, especially as similar-looking species are found across a wide range of genera. However, a good knowledge of hoverfly genus identification characters, as presented here, should enable identification of any hoverfly bumblebee mimic encountered.

WING loop in vein R_{4+5} present

LEGS hind legs all-black, triangular projection on femur	LEGS hind legs partly pale, no projection on femur

Merodon equestris pp. 29, 248
4–7 colour forms are recognised but the wing and hind leg features are consistent in every form.

Eristalis intricaria p. 228
Sexually dimorphic: males are dark with a reddish-brown 'tail'; females are somewhat larger and have a white 'tail'.
Beware – a few can look more like 'typical' *Eristalis* (see page 220).

WING loop in wing vein R_{4+5} absent   1/2

ARISTA plumose

WING upper outer cross-vein re-entrant	WING 'normal', upper outer cross-vein not re-entrant

Volucella bombylans pp. 31, 270
3 colour forms: a black form with a red-orange 'tail' resembling *Bombus lapidarius*; a black-and-yellow form resembling *B. lucorum*; and an entirely buff-coloured form, resembling *B. pascuorum*. The wing features are consistent across the forms.

Sericomyia superbiens p. 266
Upland and western distribution.
NOTE: Could be confused with the buff form of *Volucella bombylans* [check wing venation] or potentially *Criorhina floccosa* and *C. berberina* (*opposite*) [check arista].

IDENTIFYING WASP AND BEE MIMICS

| WING 'normal'; loop in wing vein R₄₊₅ absent | 2/2 |

ARISTA 'normal'; bare or pubescent

| HEAD disproportionately small in comparison to body | FACE mouth elongated downwards – diagnostic of *Criorhina* (see p.293) |

Pocota personata p.296

Criorhina ranunculi p.292

| FACE entirely yellow – unique among mimics | Several colour forms (with a red, orange or white tail) that mimic different species. |

Eriozona syrphoides p.116

Criorhina berberina p.294

Beware – can be confused with *Criorhina* that have golden-yellow dusting on the face.

FACE zygoma present – diagnostic of Cheilosiini (see *page 170*)

C. berberina (ABOVE) has two forms: one dark with a buff stripe across the thorax and a pale 'tail'; and one entirely buff-coloured that looks similar to *C. floccosa* (BELOW). The two species can usually be separated by their antenna colour.

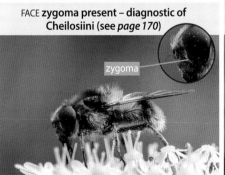

Cheilosia illustrata p.176

Criorhina floccosa p.294

A guide to the most frequently photographed hoverflies

These plates show the 36 species that are most often photographed and aims to provide a visual guide to what you are most likely to see in urban gardens, parks, *etc*. Looking at pictures alone will not always get you to a firm identification, but these plates should help to point you in the right direction if you are having trouble with the **Guide to the tribes** (see *page 53*). Some species/genera are very similar (*e.g. Epistrophe* and *Megasyrphus* are often mistaken for *Syrphus*). Coloration may be influenced by the temperature at which larvae develop (see *page 28*) and it is therefore not possible simply to rely on colour patterns. It also helps to check the distribution maps and flight time diagrams to rule out species that are unlikely to occur where and when a photo was taken! The section on **Photographing hoverflies** on *page 311* provides useful tips for obtaining good, identifiable images.

The smaller, grey images show the hoverfly's actual size. The frequency ranking of each species is shown before its name (*e.g.* **1** to **36**), but similar-looking species are shown together. The species have an 'ease of identification' colour code (● = hard; ● = care needed; ● = easy), and, where relevant, a list of species they have been confused with (! = frequently; ! = rarely).

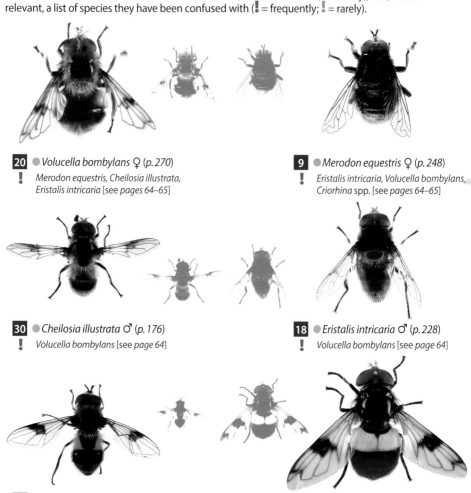

20 ● *Volucella bombylans* ♀ (*p. 270*)
! *Merodon equestris, Cheilosia illustrata, Eristalis intricaria* [see *pages 64–65*]

9 ● *Merodon equestris* ♀ (*p. 248*)
! *Eristalis intricaria, Volucella bombylans, Criorhina* spp. [see *pages 64–65*]

30 ● *Cheilosia illustrata* ♂ (*p. 176*)
! *Volucella bombylans* [see *page 64*]

18 ● *Eristalis intricaria* ♂ (*p. 228*)
! *Volucella bombylans* [see *page 64*]

29 ● *Leucozona lucorum* ♂ (*p. 116*)
! *Volucella pellucens* [*p. 272*]

8 ● *Volucella pellucens* ♀ (*p. 272*)
! *Leucozona lucorum* [*p. 116*]

MOST FREQUENTLY PHOTOGRAPHED HOVERFLIES

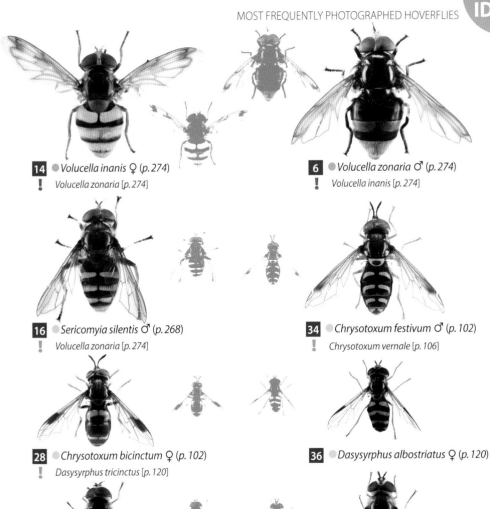

14 ● *Volucella inanis* ♀ (p. 274)
! *Volucella zonaria* [p. 274]

6 ● *Volucella zonaria* ♂ (p. 274)
! *Volucella inanis* [p. 274]

16 ● *Sericomyia silentis* ♂ (p. 268)
! *Volucella zonaria* [p. 274]

34 ● *Chrysotoxum festivum* ♂ (p. 102)
! *Chrysotoxum vernale* [p. 106]

28 ● *Chrysotoxum bicinctum* ♀ (p. 102)
! *Dasysyrphus tricinctus* [p. 120]

36 ● *Dasysyrphus albostriatus* ♀ (p. 120)

25 ● *Syrphus* sp. ♂ (pp. 132–135)
! Other *Syrphus, Eupeodes* spp. [see pages 128–129]
Epistrophe grossulariae [p. 138]

33 ● *Epistrophe grossulariae* ♀ (p. 138)
! *Syrphus ribesii* [see pages 100, 128–129]

27 ● *Meliscaeva auricollis* ♀ (p. 160)
! *Platycheirus scutatus* [p. 88]

1 ● *Episyrphus balteatus* ♂ (p. 162)

ID MOST FREQUENTLY PHOTOGRAPHED HOVERFLIES

13 ● *Eupeodes luniger* ♂ (p. 147)
! Other *Eupeodes* [pp. 148–151]
! *Scaeva pyrastri* [p. 126]

11 ● *Scaeva pyrastri* ♂ (p. 126)
! *Eupeodes luniger* [p. 147]

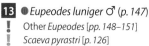

15 ● *Eupeodes corollae* ♂ (p. 118)
! Other *Eupeodes* [pp. 147–151]

19 ● *Xanthogramma pedissequum* ♀ (p. 110)
! *Xanthogramma stackelbergi* [p. 110]

7 ● *Melanostoma scalare* ♀ (p. 94)
! Other *Melanostoma* [p. 94],
some *Platycheirus* [p. 75]

35 ● *Leucozona glaucia* ♂ (p. 118)
! *Leucozona laternaria* [p. 118]

23 ● *Epistrophe eligans* ♂ (p. 136)
! *Eristalis* spp. [p. 220]

22 ● *Platycheirus albimanus* (p. 84)
! *Platycheirus scutatus* [p. 88]

17 ● *Sphaerophoria scripta* ♂ (p. 114)
! Other *Sphaerophoria* [p. 112]

10 ● *Syritta pipiens* ♂ (p. 290)
! *Tropidia scita* [p. 290]

MOST FREQUENTLY PHOTOGRAPHED HOVERFLIES ID

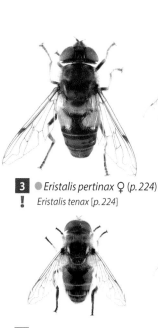

3 ● *Eristalis pertinax* ♀ (p. 224)
! *Eristalis tenax* [p. 224]

ERISTALIS SPP.
! Other *Eristalis* [p. 220]
! *Epistrophe eligans* [p. 136]

4 ● *Eristalis tenax* ♂ (p. 224)
! *Eristalis pertinax* [p. 224]

24 ● *Eristalis arbustorum* ♂ (p. 232)

32 ● *Eristalis nemorum* ♀ (p. 226)

31 ● *Eristalis horticola* ♂ (p. 230)

5 ● *Myathropa florea* ♀ (p. 236)

HELOPHILUS SPP.
! Other *Helophilus* [p. 242]
! *Parhelophilus* spp. [p. 238]

2 ● *Helophilus pendulus* ♀ (p. 242)

21 ● *Helophilus trivittatus* ♂ (p. 244)

26 ● *Xylota segnis* ♂ (p. 288)

12 ● *Rhingia campestris* ♂ (p. 194)
! *Rhingia rostrata* [p. 194]

Introduction to the species accounts

ACCOUNT ORDER
The species accounts are arranged by sub-family and tribe as follows: Syrphinae (3 tribes), Eristalinae (10 tribes) and Microdontinae (1 genus). There is an introductory guide to each tribe that highlights the key identification features and presents a key to its genera. See also **Guide to the tribes** on *page 53*.

GENERA
A brief introduction to the genus includes the main distinguishing features and a general statement about larval biology. Genera with numerous species may have a table of 'pointers' to assist in identification to species.

SPECIES
Species are organised so that similar-looking species appear as close together as practicable or in alphabetical order. **An index of genera is given on the back flap of the cover.**

SPECIES ACCOUNTS
Each account is structured in a consistent way to highlight key identification features, indicate similar species, and provide observation tips, as explained *below*:

The icons for each species indicate how straightforward a species is to identify:

 Can be identified in the field (with experience).

 Identifiable in the field by close examination by capturing the hoverfly and using a hand lens.

 Requires examination of a dead specimen under a microscope.

The camera icons indicate how straightforward it is to identify the species from photographs:

 Identifiable in the majority of cases from good photographs.

 Identification often possible using a suite of good photographs (top-down, side-on and head-on).

 Identification unlikely to be possible from photographs (microscopy required).

In tables, species that are illustrated are highlighted in **bold italics**; species that are not illustrated are in *italics* (red if threatened or rare). Vagrant species are shown in blue text.

Red List status: For Europe (TOP) and Britain (BOTTOM) (see *page 302*)

Identification icons: (see *above*)

GB/UK status: Nationally scarce/UKBAP (see *page 302*)

NT
LC
Scientific name

Wing length: range in mm (thick bar = min.; thin bar = max.); shown at actual size as scale bar (LEFT).

Identification: Brief summary of key identification features.

Similar species: Species with which confusion is most likely; those not illustrated are coded (N/I).

Observation tips: Behavioural features relevant to identification and comments on distribution or records in Britain and Ireland.

Nationally Scarce
UKBAP Priority Species
GB: Widespread ▲
Ir: Frequent

Frequency: Based on recording scheme data for Britain (GB) and Ireland (Ir) (see *opposite* for explanation). For GB, the trend is also given:
▲ increasing
▼ decreasing
= no significant change

Distribution map: Based on recording scheme data (see *opposite* for explanation):
■ most records
▨ some–few records
□ no records

MAIN IMAGES – Bearing in mind the challenges of photos and perspective, an attempt has been made to present the species to scale – the approximate scale used is shown on the plate.

Flight-period: Frequency of records each week in Britain (Jan 2000 to Sep 2022): the darkest shading represents the peak flight-period(s)

J F M A M J J A S O N D

THE SPECIES ACCOUNTS

DISTRIBUTION MAPS

Earlier editions of this book covered Great Britain and the Isle of Man, but in preparing this edition the opportunity has been taken to expand the coverage to include Ireland. The maps are based on records from 2000 onwards and have been prepared using the computer program Frescalo (from 'FREquency SCAling LOcal') (Hill, 2011) applied to combined data from the Hoverfly Recording Scheme and data supplied by the National Biodiversity Data Centre in County Waterford, Ireland, and CEDaR in Northern Ireland (see *page 323*). This analysis provides an estimation of the relative frequency with which a species has been recorded in a given hectad (10 × 10 km grid square) based on its frequency in a 'neighbourhood' – a set of around 50 hectads that are both nearby and similar in their environmental characteristics. The colour coding used is indicated in the box *opposite*.

The biggest issue associated with preparing these maps is compensating for the lack of information from poorly recorded areas. In Britain, this problem is most acute in remote areas where there are few roads and only irregular visits by hoverfly recorders – notably northern Scotland and some of the distant islands. Nevertheless, the overall range and variation shown on the maps seems to be broadly consistent with the known distribution of species.

It is not possible to be as confident about the distributions in Ireland, however, since the numbers of available records are much lower. Although certain species appear to be absent from areas where they would be expected to be present, evaluating the maps does indicate some consistent patterns of distribution. This suggests that the maps are actually quite realistic and that species' absences are related to local factors such as a lack of habitat. A much higher level of hoverfly recording in the more remote areas of Britain and Ireland would provide a clearer picture of the true status and distribution of many species. There is currently no independent hoverfly recording scheme for Ireland.

Reference

Hill, M.O. 2011. Local frequency as a key to interpreting species occurrence data when recording effort is not known. *Methods in Ecology and Evolution*, **3**: 195–205.

FREQUENCY

The terms included in the brown boxes above the maps indicate the species' frequency for both Great Britain and the Isle of Man, and for Ireland. These are based on the number of hectads of the Ordnance Survey National Grids of Great Britain and Ireland from which the species was recorded from 2000 onwards, as summarised in the following table. The frequency for species that are not the subject of a species account is included in the *List of British and Irish hoverflies* (see *page 302*).

Term	Great Britain & Isle of Man	Ireland
Widespread	> 500 hectads	> 96 hectads
Frequent	121–500 hectads	34–95 hectads
Local	41–120 hectads	9–33 hectads
Scarce	16–40 hectads	6–8 hectads
Rare	1–15 hectads	1–5 hectads

FLIGHT-PERIOD

The flight-period (phenology) chart below each map is based on the number of records each week in Great Britain from January 2000 to September 2022: the darkest shading represents the most records, and hence indicates the peak flight-period(s). Irish data were not used when preparing these charts as, in many cases, they were insufficient to generate meaningful results.

BACCHINI

from 8a p. 60

Guide to Bacchini

Although there are only four genera in the Bacchini, one of them (*Platycheirus*) is our second largest and its species can be difficult to identify. **The tribe is characterised by the absence of hairs on the humeri, a somewhat concave head which closely fits the thorax, a black face, a black scutellum** and markings on the abdomen which are usually yellow or orange, but can consist of silvery-grey or bronzy-coloured dust spots. Occasional all-black 'melanics' are found and these may cause confusion with the **Cheilosiini** (*p. 168*) or **Chrysogastrini** (*p. 196*) which also have a black face but are generally broader-bodied.

1

a) **Small; abdomen extremely elongate, narrow and waisted**

Distinctive – apart from the genera *Sphegina* (p. 198) and *Neoascia* (p. 200) [both Chrysogastrini], which are somewhat smaller and, whilst they have narrow, 'waisted' abdomens, are not so elongate.
▶ *Baccha elongata* p. 74

Baccha elongata

b) **Large; oval-bodied; abdomen broad; scutellum black; distinctive markings on abdomen**
▶ *Xanthandrus comtus* p. 92

c) **Narrow-bodied; abdomen not waisted or broad**
▶ 2

Xanthandrus comtus

2 *from 1c*

a) **FEMALES:
eyes separated at top of head**
▶ 3

b) **MALES:
eyes touching at top of head**
▶ 4

BACCHINI

3 FEMALES
a) Abdomen with distinctive triangular markings on abdomen segments T3 + T4

▶ *Melanostoma* p. 94

FEMALES
b) Abdominal markings not triangular
▶ *Platycheirus* pp. 75–93

Note: Two species have abdominal markings that are characteristic (see p. 76): *Platycheirus granditarsus* (p. 82) [small spots on T3] and *P. rosarum* (p. 82) [T2–T4] extensively orange].

4 MALES
a) Front legs not modified
▶ *Melanostoma* p. 94
▶ *Platycheirus rosarum* p. 82
▶ *Platycheirus ambiguus* p. 84
▶ *Platycheirus sticticus* (N/I) p. 80

See front leg modifications note below.

MALES
b) Front legs modified in some way
See front leg modifications note below.
▶ *Platycheirus* pp. 75–93

♂ *Platycheirus* front leg modifications

Most ♂ *Platycheirus* have modified front legs, the nature of which and where any hairs are located are useful in identification. However, both *P. rosarum* (p. 82) and *P. ambiguus* (p. 84) have front legs without any tarsal modification, possibly leading to confusion with *Melanostoma* (p. 94).

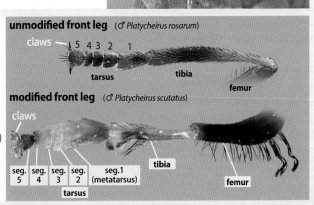

BACCHINI: *BACCHA*

Baccha
from 1a, p. 72

1 British and Irish species (illustrated)

Small, dark and with a very elongate, wasp-waisted abdomen. The larvae feed on a variety of aphids, especially nettle and bramble aphids.

LC *Baccha elongata*

LC

GB: Widespread ▽
Ir: Widespread

Wing length: 4·0–8·25 mm

Identification: The very long, slender, wasp-waisted shape and wings with the outer cross-vein joining R_{4+5} at a shallow angle (see inset *below*) should make this small hoverfly unmistakable.

Similar species: Only *Neoascia* (*p. 200*) and *Sphegina* (*p. 198*) (both Chrysogastrini) are anything like *Baccha* in shape, although those two genera have swollen hind femora and a shorter, thicker abdomen. In addition, the outer cross-veins are sharply upturned and join R_{4+5} at right-angles (see inset *below*). The wing venations of *Neoascia* and *Sphegina* are illustrated in 3b of the the *Guide to the tribes* on page 55.

Observation tips: Most frequently seen in dappled shade in woodland, along hedgerows and similar places where it manoeuvres low down among the vegetation. An occasional flower visitor that will also bask on sunlit leaves. Although widespread and common across most of Britain and Ireland, it is inconspicuous and often overlooked.

J F M A M J J A S O N D

Baccha elongata ♀ × 10

♂ × 5

The long, narrow abdomen segments 2–4 make this species unmistakable.

Baccha

Neoascia

Wing outer cross-veins compared: *Baccha* (TOP) and *Neoascia/Sphegina* (ABOVE)

BACCHINI: *Baccha* | *Platycheirus*

Platycheirus 25 British and Irish species (11 illustrated)

Small, long-and-narrow hoverflies with yellow or silvery (in a few species) abdominal markings, a black face and a black scutellum. The males of many species have modified front legs with broad, flattened tarsi. *Platycheirus* is the second largest genus of British hoverflies. Some species are readily identifiable but there are many that are difficult, especially among females, and reference to a specialist key (*e.g.* Stubbs & Falk, 2002) is required to attempt any identification of these species. *Platycheirus* can be abundant, but as the species look very similar in the field, it is necessary to retain plenty of specimens if you want to maximise your chances of finding all the species that occur at a site. The larvae feed on aphids and although some are specific to a few prey species, most seem to be general predators of leaf-litter and ground-layer aphids.

Platycheirus **identification features**

- face-shape (♂&♀) and dusting (♀)
- antenna colour
- femoral hair (♂)
- tarsal modifications (♂)
- abdominal markings and segment-shape

Features of the front leg useful in identification of males

claws | tarsus | tibia | femur

seg. 5 | seg. 4 | seg. 3 | seg. 2 | seg. 1 (metatarsus)

The shape and colour of the five tarsal segments, particularly the metatarsus

The shape and thickness of the tibia and its hair patterns

The shape, length, colour and density of hairs on the femur

Guide to the groups of British and Irish *Platycheirus* species

Although *Platycheirus* includes a few species that are readily recognised, the majority require the microscopic examination of characters for confident identification. Sexual dimorphism is noticeable, with males having many more useful identification features than females. In a few cases, it is not possible to identify females reliably and therefore wisest not to try without reference to voucher specimens. The situation is further complicated as recent research has identified several species in what was once thought to be a single species.

The approach taken here is an attempt to de-mystify the genus by creating 'morpho-groups', each with a lead species, as has previously been done for *Cheilosia*. Although these groupings are somewhat contrived it does help simplify what would otherwise be a very difficult set of species to identify.

The following is a step-by-step guide to the separation of the group lead species supported by a table (*p. 80*) with further information on the identifcation of species within the groups.

It is essential to follow the steps and to not skip any stage

Finding *Platycheirus*
Platycheirus females are very fond of Ribwort Plantain *Plantago lanceolata* pollen and are often found in numbers at this species on roadside verges. In these situations it is often productive to sweep plantain flowers, as many as five or six species can be found in one session. These may include *P. clypeatus*, illustrated here.

1 Male or female?

a) Eyes **separated** at top of head
FEMALES ▶ 2

a) Eyes **touching** at top of head
MALES ▶ 5

FEMALES (eyes separated at top of head)

2 ABDOMEN with characteristic markings (subgenus *Pyrophaena*) NOTE: male patterns similar

ABDOMEN segments T2 to T4 wholly or partially orange	ABDOMEN with pale yellow spots on segment T3 and some with markings on the margins of T4

▶ *P. granditarsus* p. 82	▶ *P. rosarum* p. 82

3 | ABDOMEN with silvery markings – either spots or bands

ALBIMANUS group [4 spp.]
▶ **p. 84** | p. 80

♀ **Platycheirus identification** Of the 'typically' yellow- or whitish-patterned ♀ *Platycheirus*, the coloration of the front femora and antenna are useful for further differentiation.

FRONT FEMORA — dark | orange/yellow

ANTENNAE — partly pale beneath | black

4 | ABDOMEN with yellow or whitish markings

4a) FRONT FEMORA partly dark

ABDOMEN SEGS. with yellow markings on at least T2 to T4

MANICATUS group [3 spp.] ▶ **p. 86** | p. 80

4b) FRONT FEMORA almost completely orange/yellow

ANTENNAE partly pale beneath

ABDOMEN SEGS. T2 with small semicircular spots positioned parallel to the segment rear margin	ABDOMEN SEGS. T2 with larger markings positioned obliquely to the segment rear margin
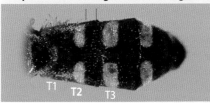	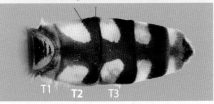
SCUTATUS group [3 spp.] ▶ **p. 88**	p. 80	PELTATUS group [3 spp.] ▶ **p. 88**	p. 80

ANTENNAE black

ABDOMEN SEGS. sides of T5 & T6 almost entirely orange	ABDOMEN SEGS. sides of T5 & T6 substantially black
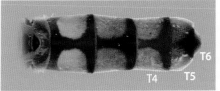	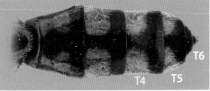
FULVIVENTRIS group [3 spp.] ▶ **p. 92**	p. 80	CLYPEATUS group [7 spp.] ▶ **p. 90**	p. 80

MALES (eyes touching at top of head)

5 ABDOMEN with characteristic markings (subgenus *Pyrophaena*) NOTE: female patterns similar

ABDOMEN segments T3 wholly orange, T2 + T4 partially orange; FRONT TARSI SEG. 1 'boxing-glove'-shape

▶ *P. granditarsus* p. 82

ABDOMEN with pale yellow spots on segment T3 and some with markings on the margins of T4; FRONT TARSI not modified

▶ *P. rosarum* p. 82

6 FRONT FEMORA with tangle of hairs at base

7 ABDOMEN with silvery/bronzy markings; FRONT TIBIAE unmodified

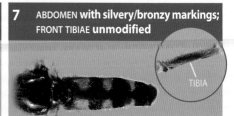

ALBIMANUS group [3 of 4 spp.]
▶ p. 84 | p. 80

ANTENNAE partly pale; **ABDOMEN** with silvery/bronzy markings; **FRONT TARSI SEGS.** 2–4 broad – almost square at apex

ALBIMANUS group [1 of 4 spp.]
▶ *P. albimanus* p. 84 | p. 80

ANTENNAE partly pale, **ABDOMEN** yellow markings; **FRONT TARSI SEGS.** 2–4 much wider than long

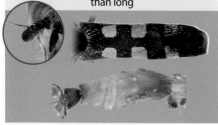

SCUTATUS group [3 spp.]
▶ p. 88 | p. 80

FRONT TARSI SEGS. unmodified

FRONT FEMUR with curled hair at apex ▶ *P. ambiguus* p. 84

FRONT FEMUR lacking curled hair at apex ▶ *P. sticticus* rare (NO ACCOUNT)

FRONT TARSI SEGS. 1–2 large + rounded; 3–5 narrow

▶ *P. discimanus* rare (NO ACCOUNT)

BACCHINI: *Platycheirus*

P. albimanus* and *P. scutatus | The vast majority of *Platycheirus* males encountered will be either *P. albimanus* or *P. scutatus* (recognisable from the femoral hair tangles, abdomen colour and the shape of the front tarsis); other species generally favour more humid/moist situations and are often best sought by sweeping (see *p. 317*).

3 ABDOMEN with **yellowish or whitish spots**; FRONT FEMORA **no tangle of hairs**

ANTENNAE black; FRONT TIBIAE almost cylindrical; FRONT TARSI SEGS. **1–2 expanded**, SEGS. **3–5 taper down**

ANTENNAE partly pale beneath; FRONT TARSI SEG. **1 'shield'-like**

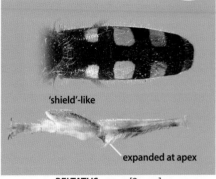

expanded — almost cylindrical

'shield'-like — expanded at apex

***MANICATUS* group** [3 spp.]
▶ *p. 86* | *p. 80*

***PELTATUS* group** [3 spp.]
▶ *p. 88* | *p. 80*

ANTENNAE black; ABDOMEN SEGS. with **narrow black margins**, T5 mostly **yellow**

ANTENNAE black; ABDOMEN SEGS. with **broad black margins**, T5 mostly **black**

narrow margins — mostly yellow — T3 T4 T5

broad margins — mostly black — T3 T4 T5

***FULVIVENTRIS* group** [3 spp.]
▶ *p. 92* | *p. 80*

***CLYPEATUS* group** [7 spp.]
▶ *p. 90* | *p. 80*

Platycheirus species in their groups:

Distinctive species (subgenus *Pyrophaena*)	
P. rosarum p. 82	**Both sexes** | pair of pale spots on abdomen segment T3.
P. granditarsus p. 82 NT	**Both sexes** | abdomen segments T2–T4 extensively or completely reddish-orange.

ALBIMANUS group	
P. albimanus p. 84	**Males** | separated by the hairs on the femur together with the shapes of the front tarsi (see *page 78*). **Females** | separated by the shape of the markings on abdomen segment T3. *P. discimanus* flies early in the spring and visits *Salix* catkins, usually flying very high up. *P. sticticus* has a very widely dispersed distribution. However, both are rarely seen.
P. ambiguus p. 84	
P. discimanus p. 85 NS	
P. sticticus NS	

CLYPEATUS group	
P. clypeatus p. 90	**Males** | separated by the shape of pits on the underside of the front metatarsus, the shape of the front tibiae, and the arrangement of bristles on the front femora. **Females** | separated by the distribution and length of hairs on the front femora, frons dusting, shapes of abdomen segment markings, and the extent of yellow on the hind femora. Most are mainly northern and western in distribution, although *P. europaeus* and *P. occultus* do occur into south-east England.
P. angustatus p. 90	
P. europaeus	
P. occultus	
P. podagratus	
P. ramsarensis	
P. scambus	

FULVIVENTRIS group	
P. fulviventris p. 92	**Males** | separated by the arrangement of bristles and hairs on the front femora and on the colour of the front femora (in combination). **Females** | *P. fulviventris* has more sharply defined edges to the frons dusting than both *P. immarginatus* and *P. perpallidus*, which cannot be separated reliably. *P. immarginatus* and *P. perpallidus* are associated with sedge beds and emergent vegetation *P. immarginatus* is almost entirely coastal, whereas *P. perpallidus* occurs inland, mainly in association with still or slow-moving water.
P. immarginatus NS	
P. perpallidus NS	

MANICATUS group	
P. manicatus p. 86	**Males** | separated by the shape and coloration of the front tarsi. **Females** | In *P. manicatus*, the frons is wholly dusted whereas in *P. tarsalis* it is partially dusted; it is completely free of dusting in *P. melanopsis*. *P. melanopsis* is a montane species that is known almost exclusively from Scotland but with old records from the Lake District, Cumbria.
P. tarsalis p. 86	
P. melanopsis NT	

PELTATUS group	
P. peltatus agg. p. 88	**Males** | separated by the arrangement of hairs on the middle tibia. **Females** | separated from *P. peltatus* by the shape of the markings on abdomen segments T2 and T3 (very unreliable) and from each other by the extent of dusting on the frons (difficult). Both species are northern and western *P. nielseni* is widespread but *P. amplus* is rarely seen.
P. amplus NT	
P. nielseni	

SCUTATUS group	
P. scutatus p. 88	**Males** | *P. scutatus* has a distinctly grey-dusted frons. *P. aurolateralis* and *P. splendidus* both have a black undusted frons; they are separated from each other by the shape and arrangement of hairs on the mid-tibia (difficult). **Females** | cannot be separated from each other and should be recorded as *P. scutatus* agg. Both *P. aurolateralis* (rarely found) and *P. splendidus* are widely distributed but are both much scarcer than *P. scutatus*.
P. aurolateralis	
P. splendidus	

Finding *Platycheirus* and *Melanostoma* larvae

Bearing in mind how numerous adults of the commoner *Platycheirus* and *Melanostoma* species can be, we know surprisingly little about their larvae – a few can be found quite readily but for many it is a challenge. There is a lot of scope for new discoveries by the diligent searcher. As a starting point, here are some examples of how to find their larvae.

The bright green larvae of *Melanostoma scalare* are comparatively easy to find in the autumn. Just collect up a bag of fallen leaves (especially those of Sycamore), take them home and work through them in comfort. It sounds simple, doesn't it? In reality, as always, diligence is the key, but these larvae can be found and reared through to maturity because they happily feed on almost any soft-bodied parcel of insect protein presented to them, especially the larvae of other flies.

Melanostoma scalare larvae are bright green and relatively easy to find.

Finding *Platycheirus* larvae is a bit more of a challenge as they feed on aphids and some are highly specialised. The clue is to watch what the adults do. One of the easier species to find is *Platycheirus ambiguus*, the males of which are often numerous, hovering close to Blackthorn. If you see the males, it is pretty likely that females will be nearby. If you return to the same Blackthorn patch a month or two later and check leaves curled by aphid infestation, there is a good chance of finding the larvae of *P. ambiguus*. Similarly, one can look for species such as *P. scutatus* and *P. splendidus* in leaf curls on various species of tree and also in aphid colonies on Hogweed and campions.

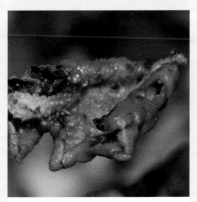

Platycheirus ambiguus larvae in a Blackthorn leaf curled by aphids.

There are some *Platycheirus* species, such as *P. podagratus*, about which we know precious little. *P. immarginatus* seems to favour coastal wetlands, and in the Spey Valley, sedge beds are often teeming with various *Platycheirus* species so we can assume that there is a good chance of finding larvae associated with sedge aphids. There is so much to learn and if you have the ability and patience, diligent searching stands a good chance of unlocking new biological information that may help to conserve more habitat-specific species.

Platycheirus albimanus larvae in a Hogweed sheath.

Subgenus *Pyrophaena* — 2 British and Irish species (illustrated)

At present, the British and Irish list includes two species of *Platycheirus* in the subgenus *Pyrophaena*. In Europe, *Pyrophaena* is treated as a separate genus and the taxonomic arrangement of the species concerned in Britain and Ireland may change in future.

Platycheirus granditarsus

NT | **LC**

GB: Widespread =
Ir: Widespread

Wing length: 5·25–8·5 mm

Identification: A readily recognised species in which both sexes have large orange-red markings on the abdomen. In some females, the abdominal markings are limited, making them more difficult to recognise. The male has both the front and middle legs modified with black peg-like extensions to the tarsi.

Similar species: Possibly superficially confused with *Xylota segnis* (*p. 288*), which also has orange-red markings on the abdomen.

Observation tips: A widespread and often abundant wetland hoverfly which occurs in wet meadows and along the edges of water bodies such as ponds and ditches. Whilst widespread across Britain and Ireland, this species is less frequently encountered in south-east England where there are indications that it is disappearing from some traditional haunts.

J F M A M J J A S O N D

Platycheirus rosarum

LC | **LC**

GB: Widespread =
Frequent

Wing length: 5·25–7·75 mm

Identification: Not immediately recognisable as a *Platycheirus* because there are no obvious modifications to the legs of the male. However, both sexes are readily recognised by the unusual shape and markings of the abdomen which widens steadily towards the tip, and has a pair of pale yellow spots on abdomen segment T3 (and in some individuals on T4) on an otherwise black background. When the wings are folded over the abdomen, the distinctive spots on the abdomen are often difficult to see owing to the strong dark suffusion on the wings.

Similar species: None.

Observation tips: A widespread, but local species of wet meadows and along the edges of water bodies such as ponds and ditches. Often abundant where it occurs.

J F M A M J J A S O N D

ALBIMANUS group
4 SPP. | 2 ILL.

Platycheirus albimanus
GB: Widespread =
Ir: Widespread

Wing length: 5–8 mm

Identification: Males have dark metallic bronze or silvery-grey markings on the abdomen and very distinctive front legs with the front femora having a clump of tangled hairs near the base, and modified tarsi, with front tarsi seg. 2 being almost square. Females are readily identifiable by their grey-spotted black abdomen and distinctive dust spots on the frons.

Similar species: See box *below*; also *Platycheirus scutatus* (*p. 88*) and related species in the SCUTATUS group (see *page 80*), the ♂'s of which have similar hair clumps at the base of the front femur, but the front tarsal segments are much narrower.

Observation tips: Widespread, abundant and found throughout the year, but perhaps most numerous in the spring. It occurs among low foliage, such as Bramble and nettle patches, and at a very wide range of low-growing flowers. Common in gardens.

J F M A M J J A S O N D

Similar species: Both *Platycheirus albimanus* and *P. ambiguus* can be confused, especially as they often occur together around Blackthorn. Compared with *P. albimanus*, ♂ *P. ambiguus* are somewhat smaller, more delicate (which can be noticeable when they fly together) and the front tarsi are unmodified (see *opposite* and the species accounts). In addition, the frons of *P. ambiguus* is strongly dusted and broader. The Nationally Scarce and rarely reported *P. discimanus* (N/I) and *P. sticticus* (N/I) may also occur with these species, and differ as follows: ♂ *P. discimanus* front tarsal modifications (see *opposite*); ♂ *P. sticticus* front tarsi unmodified and lacking femoral curled bristle. Confusion with other genera is unlikely, but occasionally reports of *Portevinia maculata* (*p. 192*) have proven to be female *P. albimanus* [*Portevinia* has orange antennae].

Platycheirus ambiguus
GB: Frequent ▽
Ir: Local

Wing length: 4·5–7·0 mm

Identification: Easily overlooked among other *Platycheirus*. Males have unmodified front legs and a very distinctive curious curled bristle near the apex of the femur (although this may be visible on close examination using a hand lens, experience is required to ensure the correct angle of view). The frons is broad and strongly silvery-dusted. Females have grey bands on abdomen segments T3 and T4 which can be quite faint and only visible from oblique angles.

Observation tips: Males often hover close to flowering Blackthorn, sometimes in small swarms. The larvae can be found feeding on aphids on Blackthorn and possibly other *Prunus* species. Most frequent in south-east England; there are very few records from Ireland.

J F M A M J J A S O N D

BACCHINI: *Platycheirus*

Platycheirus albimanus × **8**

♂

♀ | characteristic narrow dust spots on the frons

tangle of hairs at base of front femur — tibia

ALBIMANUS GROUP
♂ **MODIFIED FRONT TARSI**

P. discimanus
no hair tangle
SEGS. 1–2 large and rounded; 3–5 narrow

P. albimanus
hair tangle
SEG. 2 almost square

NOTE: Both *P. ambiguus* and *P. sticticus* have unmodified front tarsi.

♂ *Platycheirus ambiguus* × **8**

curled bristle at apex of front femur — tibia

The tibia and tarsi of ♂ *P. ambiguus* are unmodified.

MANICATUS group 3 SPP. | 2 ILL.

Platycheirus manicatus

GB: Widespread ▽
Ir: Frequent

Wing length: 6·75–9·0 mm

Identification: Both sexes have the thorax and head extensively dusted giving them a dull, slightly bronzy appearance. The face is extended forwards and dusted. Males have front legs with tarsal segments 3–5 all the same mid-grey-brown (or darker). Females have front femora strongly darkened at least in the basal half.

Similar species: Other MANICATUS group species (see *page 80*). Female *P. peltatus* (*p. 88*) could cause confusion but they have wholly orange front femora.

Observation tips: A dry grassland species which often occurs on calcareous soils, coastal grasslands and moorlands. It can be one of the commonest species in exposed and open habitats in the north and west of Scotland and the Northern Isles. Records suggest that this species is not present in more northerly parts of Ireland but this may be an artifact of recorder effort.

J F M A M J J A S O N D

Platycheirus tarsalis

GB: Frequent ▽
Ir: Not recorded

Wing length: 7·5–8·75 mm

Identification: Both sexes have a shiny black thorax (rather than dull and dusted). The face is extended forwards, but less so than in *P. manicatus*, and dusted. Males have front legs with tarsal segments 3–5 that are typically quite dark, with the 5th segment paler and contrasting. Females have front femora strongly darkened at least in the basal half.

Similar species: Other MANICATUS group species (see *page 80*). Confusion with *P. peltatus* (*p. 88*) is possible from photographs if they do not show the tarsal segments [♂] or the front femora [♀].

Observation tips: A woodland species that is usually found in open spaces such as rides and clearings, where it is a regular visitor to Hedge Mustard. Although widespread, it is most abundant in the midlands and the Welsh border counties. It flies mainly in May and June. Records outside this date range should be treated with considerable caution as they are more likely to refer to *P. peltatus*.

J F M A M J J A S O N D

BACCHINI: *Platycheirus*

MANICATUS group identification pointers

SPECIES	THORAX	♂ FRONT TARSI	♀
P. manicatus	dull	SEGS. 1–2 wide, pale; SEGS. 3–5 narrower; all the same colour	FRONS dusting extends to and behind ocellar triangle
P. tarsalis	shiny	SEGS. 1–2 broad, pale; SEGS. 3–5 narrower; quite dark; SEG. 5 paler than SEGS. 3–4	FRONS dusting does not reach or go behind ocellar triangle
P. melanopsis		SEGS. 1–2 broad; SEG. 3 relatively broad; SEGS. 4–5 narrow; dark	FRONS mostly shiny; dust confined to margins

Note The rare *P. melanopsis* is only found high on mountains, mainly in Scotland

♂ *Platycheirus manicatus* × 8

THORAX dull and dusted

♂ & ♀ | face extended forwards; dusted

♂ | front tarsal segments 1–2 particularly wide (almost spoon-shaped); segments 3–5 all the same colour

♂ *Platycheirus tarsalis* × 8

THORAX shiny

♂ & ♀ | face less extended forwards than in *P. manicatus*; dusted

♂ | front tarsal segments 1–2 broad; segments 3–5 quite dark, with segment 5 paler than segments 3–4

SCUTATUS group
3 SPP. | 1 ILL.

Platycheirus scutatus

Wing length: 5.0–7.5 mm

Identification: Antennal segment 3 is yellow beneath in both sexes. The abdominal markings are yellow – on T2 in the female they are small, slightly crescent-shaped and set at an angle. Males have a distinctive group of tangled hairs at the base of the femur, the front metatarsus is inflated and is followed by four segments that are broad but very short, and the frons is heavily grey-dusted.

Similar species: *Platycheirus albimanus* (*p. 84*) males also have a tangled clump of hairs on the front femur, but tarsal segment 2 is almost square and the frons lacks dusting. Other SCUTATUS group species (*P. aurolateralis* (N/I) and *P. splendidus* (N/I) – see *page 80*) are very similar, but the males have a shiny black frons and different hair arrangements on the second tibia. Females (see table on page 80) cannot be separated and should be recorded as *P. scutatus* agg. Dark female *Meliscaeva auricollis* (*p. 160*) can cause confusion if the scutellum is strongly darkened.

J F M A M J J A S O N D

Observation tips: Often occurs with *P. albimanus* in woodland rides, grasslands at the interface with scrub and woodland, and in gardens. It has a long flight-period. *P. scutatus*, *P. aurolateralis* and *P. splendidus* are all present in Ireland, with *P. scutatus* the most frequently recorded.

PELTATUS group
3 SPP. | 1 ILL.

Platycheirus peltatus

Wing length: 7–9 mm

Identification: A comparatively large *Platycheirus* with antennal segment 3 orange/yellow beneath. The male front metatarsus is distinctively asymmetrical and shield-like. The middle tibiae are also modified (careful examination is essential). Females have orange front femora; **they cannot always be separated from *P. nielseni*.** Identification of both sexes from photos is rarely possible and it should be recorded as *P. peltatus* agg.

Similar species: Other PELTATUS group species – *Platycheirus amplus* (N/I) and *P. nielseni* (N/I) – are the most likely to cause confusion (see table on *page 80*). The SCUTATUS group may be confused in photographs if the basal abdominal markings are obscured by the wings. PELTATUS group species are sometimes confused with the MANICATUS group (*p. 86*) but in that group males differ in the shapes of their front metatarsi and females have black front femora.

J F M A M J J A S O N D

Observation tips: Once widespread in lowland Britain, especially in damp woodland rides and ditches, it is disappearing from south-east England. In northern Britain *P. peltatus* occurs with the very similar *P. nielseni*, which has been recorded farther south (possibly erroneously). *P. peltatus*, *P. amplus* and *P. nielseni* are all present in Ireland, with *P. peltatus* the most frequently recorded.

BACCHINI: *Platycheirus*

♂ MODIFIED FRONT TARSI

SCUTATUS GROUP (*P. scutatus*)

tarsal segments 2–4 much wider than long

clumps of tangled hair at base of femur (cf. *P. albimanus* – p. 80) and in the middle part of the tibia

PELTATUS GROUP (*P. peltatus*)

assymetric shape of tarsal segment 1 (metatarsus) very distinctive

♂ *Platycheirus scutatus* × **8**

Platycheirus peltatus × **8**

Abdominal markings set at an angle are characteristic of the *PELTATUS* group.

CLYPEATUS group   7 SPP. | 2 ILL.

Platycheirus angustatus

GB: Widespread =
Ir: Frequent

Wing length: 5–7 mm

Identification: This is a relatively straightforward species to identify under high magnification, but males cannot be identified with certainty in the field. Females have an abdomen that is noticeably tapered to a long point which is more pronounced than in other similar species. Males have distinctive lightning-shaped pits on the undersides of the front metatarsi – see *Similar species* below.

Observation tips: A widespread and fairly common species which occurs mainly in damper situations in woodlands, grasslands and wetlands. It is frequently found at plantain and sedge flowers, as are similar species such as *P. clypeatus*.

J F M A M J J A S O N D

Similar species: All members of the CLYPEATUS group (*P. clypeatus*, *P. angustatus*, *P. podagratus* (N/I), *P. europaeus* (N/I), *P. occultus* (N/I), *P. perpallidus* (N/I) and *P. ramsarensis* (N/I)) are difficult to differentiate – see table on *page 80*. Males have a pit on the underside of each front metatarsus which is only discernible when viewed at high magnification from beneath. Details of the shape and arrangement of the pit are used in identification (*e.g.* the pits of *P. angustatus*, *P. ramsarensis* and *P. europaeus* are lightning-shaped; those of *P. clypeatus* and *P. occultus* are distinctively elongated). Females are much harder to separate and differ subtly in the shape and markings of the abdominal segments. These identification complications mean that most photographs can only be identified as *P. clypeatus* agg. *Melanostoma mellinum* [♂] (*p. 94*) may also be confused but has thin, unmodified legs.

Platycheirus clypeatus

GB: Widespread =
Ir: Widespread

Wing length: 5·0–7·5 mm

Identification: This is one of a group of very similar yellow-marked species with completely black antennae which are difficult to identify. Males have distinctive elongate pits on the underside of their front metatarsi– see *Similar species above*.

Observation tips: A common and widespread grassland species which is often found at plantain, grass and sedge flowers. It occurs from April onwards but becomes more frequent by midsummer.

J F M A M J J A S O N D

BACCHINI: *Platycheirus*

♂ front tarsi underside pits

P. angustatus — lightning-shaped

P. clypeatus — round pit at the end of a pale streak

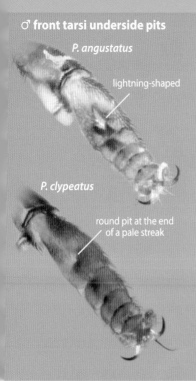

♀ *Platycheirus angustatus* × **8**

♀ | extensively black and sharply pointed abdomen

Platycheirus clypeatus × **8**   ♂   ♀

FULVIVENTRIS group
3 SPP. | 1 ILL.

Platycheirus fulviventris

GB: Frequent ▽
Ir: Frequent

Wing length: 6·0–7·5 mm

Identification: One of three *Platycheirus* species with wholly black antennae and with extensive orange markings on the abdomen (including segments T5 and T6) that give a rather colourful appearance. Males have distinctive front legs, with the front tibia very broad for most of its length (see *opposite*). Females have sharply defined dust markings on the frons (fuzzy-edged in the other two species) (see *opposite*).

Similar species: Other members of the FULVIVENTRIS group: *Platycheirus immarginatus* (N/I) and *P. perpallidus* (N/I) – see table on *page 80*. *P. scambus* (N/I) (CLYPEATUS group), which is darker.

Observation tips: A wetland species that occurs among vegetation fringing ponds, grazing marsh ditches and river banks. Its distribution across Britain and Ireland is as patchy as the wet habitat that it inhabits, with a tendency towards coastal locations and river valleys.

J F M A M J J A S O N D

Xanthandrus (1 British species illustrated)
from 1b *p. 72*

A reasonably large, robust hoverfly that has yellow-and-black abdominal markings and a completely black face and scutellum. The shape of the markings on the abdomen are quite distinctive. Larvae feed on colonial micro-moth caterpillars on a variety of trees and bushes.

Xanthandrus comtus

GB: Frequent =
Ir: Local

Wing length: 8·75–11·5 mm

Identification: Abdomen segment T2 has round spots and segments T3 and T4 have semicircular spots which may join in the middle. Males are considerably more colourful than females, which have darker and greyer markings on the abdomen.

Similar species: None.

Observation tips: Widespread but scarce; more frequent along the south coast. Regarded as a migratory species that occurs in very variable numbers each year. It is often seen during winter months at Ivy or on sunlit leaves and, like some other bacchines, can be affected by the fungus *Entomophthora muscae*. There are very few records from Ireland.

J F M A M J J A S O N D

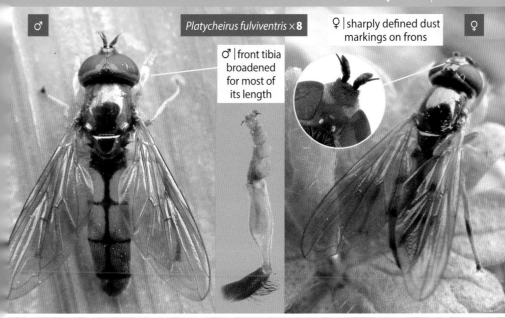

Platycheirus fulviventris × 8

♂ | front tibia broadened for most of its length

♀ | sharply defined dust markings on frons

Xanthandrus comtus × 8

from **Melanostoma** 3 British and Irish species (2 illustrated)
3a, 4b
p. 73 Small, yellow-and-black hoverflies with a completely black face and scutellum that are best found by sweeping in grassy places. **Females have distinctive triangular abdominal markings. Males can be confused with *Platycheirus* (pp. 75–93), but do not have modified front legs.** Recent DNA analysis has revised this genus, which is now believed to comprise four species (formal confirmation that all four species occur in Britain and Ireland is awaited). This analysis complicates the identification situation, as it is now difficult to be sure that characters used to identify the two commonest species are reliable. The larvae feed on a variety of aphids and other Diptera larvae among leaf-litter and the ground layer. Adults often feed on the pollen of plantains, grasses and sedges. Dead hoverflies found hanging below grass and flower heads are often *Melanostoma* that have been killed by the fungus *Entomophthora muscae*.

LC *Melanostoma mellinum*

GB: Widespread =
Ir: Widespread

Wing length: 4·75–7·0 mm

Identification: Separation from *M. scalare* requires careful examination. Females have very narrow dust spots on the mostly shining black frons. Males have a relatively short abdomen in which segments T2 and T3 are no longer than they are wide. In the uplands, there is a need for considerable caution as *Melanostoma* can be very variable and potentially comprise at least two species.

Observation tips: A widespread and abundant grassland species which can be numerous in the uplands, especially around wet flushes on moorland.

Similar species: All *Melanostoma* look similar. The most frequent confusion is with male *Platycheirus clypeatus* (p. 90) and some other yellow-marked *Platycheirus* (see pp. 75–93) but males of those species have modified front legs. *Pelecocera* (p. 252) can also be mistaken for *Melanostoma* in the field.

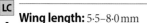
J F M A M J J A S O N D

LC *Melanostoma scalare*

GB: Widespread ▲
Ir: Widespread

Wing length: 5·5–8·0 mm

Identification: Separation from *M. mellinum* requires careful examination. Females have broad dust spots, extending most of the way across the frons. Males have an elongate abdomen in which the segments T2 and T3 are much longer than they are wide, giving them a distinctive appearance that, with experience, is relatively easy to pick out.

Observation tips: A widespread and abundant grassland species, but less frequent in the uplands. Often found in less open situations than *M. mellinum* such as woodland rides and scrubby grassland. It has a very long flight season, and is more prominent in spring than *M. mellinum*.

J F M A M J J A S O N D

BACCHINI: *Melanostoma*

♀ frons dusting

M. mellinum dust spots **narrow**

M. scalare dust spots **broad**

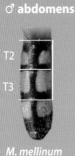

♂ abdomens

M. mellinum T2 & T3 both **as long as wide**

M. scalare T2 & T3 both **longer than wide**

♂ *Melanostoma mellinum* × **10**

♂ *Melanostoma scalare* × **10** ♀

♀ *Melanostoma* have distinctively shaped abdominal markings

***Melanostoma* species not otherwise covered:**

M. dubium	Recent DNA analysis sinks this species and erects two new ones: *M. mellarium* and *M. certuum*, both of which are likely to be added to the British list in due course.

Guide to Paragini
from 9a p.60

Represented by a single genus of tiny (usually <5 mm long) black hoverflies with at least some yellow on the face (look carefully as this can be a subtle feature). **They can be told from other similar looking species by the bare humeri, which are obscured by the head.**

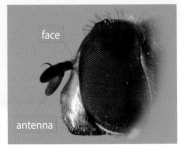

Paragus head showing the partially yellow face and elongate antennae.

Paragus   5 British and Irish species (1 illustrated)

Tiny hoverflies that visit low-growing flowers, especially yellow composites and Tormentil. They are usually black with few markings, but the legs are extensively yellow. **Microscopic examination is needed for reliable identification.** Larvae feed on aphids.

LC *Paragus haemorrhous*
LC
GB: Widespread =
Ir: Frequent

Wing length: 3·5–4·0 mm

Identification: A tiny black hoverfly that has a slightly constricted abdomen. The face is partially yellow and the antennae are somewhat elongated. It needs to be examined in detail under the microscope to separate it from others in the genus.

Similar species: Other *Paragus*, but confusion most likely with *P. constrictus* (N/I – Irish records only) and the Near Threatened *P. tibialis* (N/I), only the males of which can be identified by microscopic examination of the genitalia.

Observation tips: A grassland hoverfly that occurs in short swards in dry, sunny situations, often in the vicinity of Cats-ear. Adults rest on, or hover just above, patches of bare sand or earth along paths and banks. It is predominantly a southern species, becoming scarce in northern England and Scotland, where it is absent from higher ground.

J F M A M J J A S O N D

Paragus species not otherwise covered:

P. albifrons	CR	A species that has declined greatly: the few recent records are from the Thames marshes, generally on shingle or dry grassland close to the coast. It has a strong band of white hairs on the eyes that may be visible in good photographs (caution: there are other similar European species).
P. quadrifasciatus		Recently added to the British list, its current status is unclear. It has white hair bands on the eyes, like *P. albifrons*, and yellow markings on the abdomen.
P. constrictus	EN	Recorded only from limestone pavements in Ireland. Identification is only possible by microscopic examination of the male genitalia.
P. tibialis	NT	Modern records are restricted to heathlands in Hampshire, Surrey and Dorset. Identification is only possible by microscopic examination of the male genitalia.

PARAGINI: *Paragus*

♂ abdomens

P. haemorrhous

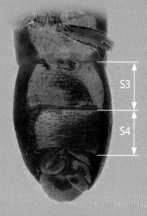

Abdomen underside segment S4 about the same length as S3.

P. tibialis (& *P. constrictus*)

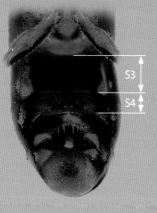

Abdomen underside segment S4 is much shorter (narrower looking) than S3.

♀ abdomens

P. haemorrhous

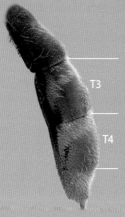

In ♀ *P. haemorrhous*, abdomen segments T3 & T4 have broad bands of black hairs.

In *P. constrictus* and *P. tibialis* these bands are very narrow.

♂ *Paragus haemorrhous* × 8

Some individuals have red markings on the abdomen.

Guide to Syrphini

from 9b p.60

A large tribe comprising 18 genera, including nearly one third of the region's species, and some of the most colourful and familiar. Most are moderate to large in size but, within species, individuals can be very variable in size depending on seasonal fluctuations in the abundance of prey items for their larvae. **The absence of hairs from the humeri, together with the concave head-shape (which fits snugly over the front of the thorax), and the presence of yellow markings on the face separate this tribe from all others apart from the** Paragini – **which, in Britain and Ireland, are all tiny, black hoverflies** – (*p. 96*). Most syrphinids have an abdominal pattern of black-and-yellow stripes or spots and most have a yellow scutellum – there are exceptions (*e.g.* some *Leucozona*), and a few have melanic or near melanic forms. Abdomen pattern can be extremely helpful in separating them, but some patterns occur in several genera whose identification is then largely reliant upon microscopic characters. **It should be remembered that members of other tribes have some features in common with the Syrphini – such as black-and-yellow abdominal patterns or yellow faces, but these all have visible, hairy humeri.**

1 a) **Sides of the thorax with distinct, sharply defined yellow markings; lacking dusting and only weakly hairy**

Beware – it is the markings on the thorax side that are meant here, not the stripes running from front to back along the top edge of the thorax surface.

▶ **2**

b) **Sides of the thorax strongly hairy or dusted, with no distinct yellow markings**

The sides of the thorax may have patches of murky yellowish or olive dusting, but not sharply defined citrus-yellow markings.

▶ **3**

2 *from 1a*

a) **Abdomen broad:**

Long, porrect antennae
▶ *Chrysotoxum* pp. 102–107

'Normal' antennae
▶ *Xanthogramma* p. 108

b) **Abdomen narrow:**

Large with darkened wings
▶ *Doros* p. 108

Small with clear wings
▶ *Sphaerophoria* pp. 112–115

SYRPHINI

3
from 1b

a) **Bumblebee mimic:**
   Scutellum and face yellow
   ▶ *Eriozona syrphoides* p. 116

b) Abdomen with large creamy white or bluish rectangular markings on the 2nd segment, much larger than any markings of the rest of the abdomen
   ▶ *Leucozona* pp. 116–119

c) Unconvincing wasp mimics; black abdomens with yellow, orange or creamy-yellow markings:
   Wing with obvious long, black stigma
   ▶ *Dasysyrphus* pp. 120–123
   Wing with less obvious, usually brown or yellow stigma ▶ **4**

4
from 3c

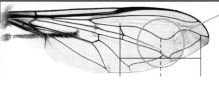

a) Wing with vein R_{4+5} distinctly dipped; basal half of discal cell parallel-sided or even narrowing
   ▶ *Didea* p. 124
   ▶ *Eupeodes lapponicus* p. 150
   ▶ *Megasyrphus erraticus* p. 131

b) Wing with vein R_{4+5} not dipped; discal cell steadily widens towards wing edge ▶ **5**

5
from 4b

To establish the identification of the 'black-and-yellow' genera that do not have a dip in vein R_{4+5} is not completely straightforward. Markings on the abdomen are helpful in some cases, but some patterns occur in several genera. These genera can only be separated reliably by features that are hard to see without strong magnification.

The table overleaf details the most apparent characteristics, as well as an indication of any reliable features for identification:

Episyrphus p. 162	*Melangyna* p. 154	*Eupeodes* p. 146
Meligramma p. 150	*Syrphus* p. 132	*Scaeva* p. 126
Epistrophe p. 136	*Meliscaeva* p. 160	*Parasyrphus* p. 142

▶ **6 – page 100**

SYRPHINI

6 Simplified key to the 'black-and-yellow/white' Syrphini genera that have wing vein R$_{4+5}$ not dipped and lack defined yellow markings on side of thorax – diagnostic features are in **bold text**.

Syrphini abdomen markings
Examples of markings and the terms used to describe their shape in the species accounts:
i) **Banded** (T2 separated)
ii) **Oval-squarish**
iii) **Lunulate**
iv) **Comma/hockey-stick/golf-club-shaped**

banded   oval-squarish   lunulate   hockey-stick

Upper surface of squamae
The squama is a membrane located where the hind margin of the wing meets the thorax. *Syrphus* has **upstanding hairs on the upper surface**; all other Syrphini genera in this section have a bare upper surface.

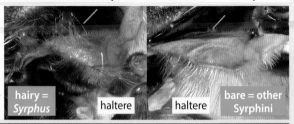

hairy = *Syrphus* | haltere | haltere | bare = other *Syrphus*

Genera that typically have all-yellow faces (see *opposite, top*)

Beware – some *Syrphus*, *Epistrophe*, *Eupeodes* and *Parasyrphus* have a very similar overall appearance. It is essential to examine all the detailed features that are indicative or diagnostic of the genera. In addition, some *Eupeodes* may have faint black marks on the lower part of the face.

hairs on upper surface of squamae	no hairs on upper surface of squamae	
ABDOMEN MARKINGS yellow; T3 & T4 banded	ABDOMEN MARKINGS yellow; T3 & T4 banded. ABDOMEN MARGINS T3 & T4 with yellow hairs	ABDOMEN MARKINGS yellow; T3 & T4 some spp. banded, some lunulate. ABDOMEN MARGINS **T3 & T4 with black hairs**

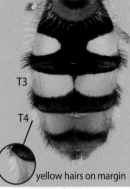
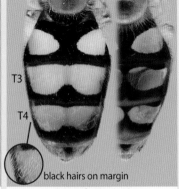

yellow hairs on margin | black hairs on margin

Syrphus [5 spp.] ▶ *p. 132*
Epistrophe [7 spp.] ▶ *p. 136*
Eupeodes [9 spp.] ▶ *p. 146*

See *page 128* for more information on the pitfalls of separating *Syrphus* (without checking detailed features) from other species that have an abdomen with a banded pattern.

SYRPHINI

Syrphini faces
Whether the face is entirely yellow (LEFT) or marked with a dark stripe (which can be partial or complete) down the centre (RIGHT) can help in identification. **Beware** – this feature can vary in intensity at both generic and specific level.

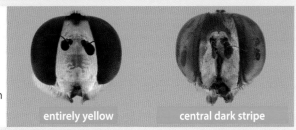

entirely yellow | central dark stripe

Genera that often have a partial or complete darkened central stripe on the face

ABDOMEN MARKINGS **yellow–cream; T2 triangular, T3 & T4 oval–squarish**. FACE 'KNOB' **yellow**.

ABDOMEN MARKINGS yellow–cream; T3 & T4 oval–squarish. FACE 'KNOB' **darkened**. FORM relatively small and narrow; LEGS mainly black.

ABDOMEN MARKINGS white or pale yellow; T3 & T4 comma-shaped. FRONS **swollen**.

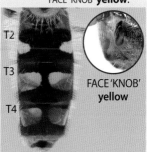

FACE 'KNOB' yellow

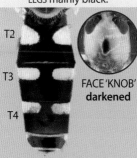

FACE 'KNOB' darkened

FRONS swollen

Meligramma [3 spp.]
▶ p. 150

Melangyna [9 spp.]
▶ p. 154

Scaeva [5 spp.]
▶ p. 126

ABDOMEN MARKINGS yellow; T3 & T4 moustache-shaped bands (except *P. punctulatus* ♂ – golf-club-shaped spots ♀ – paired spots)
Beware – some look very similar to *Syrphus* – **examine squamae** (see opposite).

Beware – face can be all-black; ♀ *M. quadrimaculata* has no markings.

ABDOMEN MARKINGS yellow; T3 & T4: *M. cinctella* – banded; *M. auricollis* – squarish spots (variable). WINGS rather elongate; **black spots on hind margin** (at high magnification)

Beware – some *Dasysyrphus* have narrow comma-shaped markings but wing stigma is long and black (see **3c**).

ABDOMEN MARKINGS **diagnostic: orange bands split into two bands by narrow black bars**.

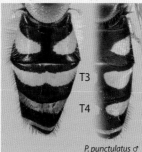

P. punctulatus ♂

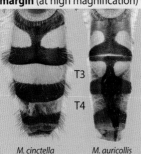

M. cinctella | M. auricollis

Parasyrphus [6 spp.]
▶ p. 142

Meliscaeva [2 spp.]
▶ p. 160

Episyrphus balteatus [1 sp.]
▶ p. 162

Chrysotoxum

from 2a, p. 98

8 British and Irish species (6 illustrated)

Large and distinctive wasp mimics with yellow markings on the side of the thorax and long, straight (porrect) antennae. Little is known about larval biology, but they are believed to feed on root aphids in ants' nests. The genus splits into three groups: *C. bicinctum*, which has very distinctive markings; *C. festivum* and *C. vernale*, in which the abdominal markings do not go all the way to the edges, leaving the margin as a continuous black slightly raised rim (beading); and the others (*C. arcuatum*, *C. cautum*, *C. elegans*, *C. octomaculatum* and *C. verralli*), which can be quite difficult to separate and are therefore referred to as the 'difficult five' – see *page 104*.

Distinctive *Chrysotoxum*

 ### *Chrysotoxum bicinctum*

GB: Widespread ▲
Ir: Widespread

Wing length: 7·0–10·25 mm

Identification: The distinctive yellow bars on abdomen segments T2 and T4 combined with the chocolate-coloured wing markings are unmistakable.

Similar species: There should be no confusion with any other hoverfly in Britain and Ireland, although *Dasysyrphus tricinctus* (*p. 120*), which lacks extensive dark wing markings and yellow markings on the sides of the thorax, might be mistakenly identified as *Chrysotoxum bicinctum* in some photographs.

Observation tips: Found in open grasslands and grassy woodland rides. Common in southern England, becoming scarcer to the north, but has been recorded as far north as the north coast of Scotland. Widely distributed in Ireland but seemingly less so in the north-east.

J F M A M J J A S O N D

'Beaded' *Chrysotoxum*

 ### *Chrysotoxum festivum*

GB: Widespread =
Ir: Local

Wing length: 8·25–12·0 mm

Identification: The distinct downturned bars on abdomen segments T2–T4 together with its size and wholly orange legs (or partially black on the front femora only) make this species quite distinctive.

Similar species: It can only be confused with the Endangered and very rare *Chrysotoxum vernale* (N/I), which is confined to very restricted parts of southern Hampshire and Dorset. The differences are subtle and involve the shape of the abdominal markings and the extent of dark markings on the legs – features which are both rather variable.

Observation tips: Predominantly a southern species in Britain, but occurs into southern Lancashire with a scattering of records north to Scotland. There are indications that this species is spreading northwards in response to climate change; it should therefore be watched for in lowland localities in the north of England and on the coast in Scotland. In Ireland, it is restricted primarily to the south-west.

J F M A M J J A S O N D

SYRPHINI: *Chrysotoxum*

WING chocolate-coloured markings

abdominal markings highly distinctive

♀ *Chrysotoxum bicinctum* ×6

♂ *Chrysotoxum festivum* ×6

yellow abdominal markings do not reach the edge – leaving a black 'bead' down side

The 'difficult 5'

Similar species: The 'difficult five' are: *Chrysotoxum arcuatum* and *C. cautum* (both *below*), *C. elegans* and *C. verralli* (*p. 106*), and the Endangered *C. octomaculatum* (N/I), which is confined to the heaths of Dorset and Surrey. Separation requires careful examination – see table *below* for a summary of the differences between these species.

SPECIES	FEATURES	
C. cautum (*below*)	ANTENNA: seg. 1 about ⅔ seg. 2; **length seg. 1 + seg. 2 ≥ seg. 3**	
C. arcuatum (*below*)	ANTENNA: seg. 1 = seg. 2; **length seg. 1 + seg. 2 < seg. 3**	
C. elegans (*p. 106*)	ANTENNA: seg. 1 = seg. 2; seg. 1 + seg. 2 > seg. 3	ABDOMEN: black bands **widen towards sides**
C. verralli (*p. 106*) *C. octomaculatum*	ANTENNA: seg. 1 = seg. 2; seg. 1 + seg. 2 ≈ seg. 3	**Note:** Separation of these species is difficult. Subtle differences in the abdomen patterns require comparison with known specimens. Records of *C. octomaculatum* require independent verification.

Chrysotoxum cautum

GB: Frequent ▽
Ir: Rare

Wing length: 10·25–13·0 mm

Identification: This is one of the biggest and bulkiest of the 'difficult five'. Both sexes can be identified on the basis of the ratio of the length of the 1st and 2nd antennal segments: segment 1 is about two-thirds of the length of segment 2 (see *opposite*). Males are easier to distinguish from other *Chrysotoxum* as they have an extremely large and obvious genital capsule.

Observation tips: A predominantly southern species which occurs as far west as Pembrokeshire in south Wales and as far north as Lincolnshire in the east. During peak emergence in late May and early June, it can be found in grasslands and open sunny rides. To date there have been only two confirmed records from Ireland.

J F M A M J J A S O N D

Chrysotoxum arcuatum

GB: Frequent ▽
Ir: Frequent

Wing length: 8·0–10·25 mm

Identification: This is the smallest member of the 'difficult five' but is relatively straightforward to identify because antennal segment 3 is considerably longer than segments 1 and 2 added together (see *opposite*). In the field, its abdomen appears to be more globular in shape than in the other species.

Observation tips: A northern and western species which is most frequently found in coniferous woodlands north of a line between the Humber and the Severn. This species was formerly recorded in north Norfolk but has not been reported this century. It was part of a northern fauna that occurred as an outlier in this area and may be a casualty of climate change. Its distribution in Ireland is somewhat patchy but it seems to be most frequent to the west.

J F M A M J J A S O N D

NT LC Chrysotoxum elegans

Nationally Scarce
GB: Local =
Ir: Not recorded

Wing length: 9·5–12·0 mm

Identification: One of the two largest of the 'difficult five' (with *Chrysotoxum cautum* (p. 104)), this is one of the trickiest to identify, requiring careful examination under magnification. The shape of the yellow bars on the abdomen, which are well separated from the front edges of the segments by a black band that widens towards the sides, is the main character used to separate it from *C. verralli* and *C. octomaculatum* (N/I) (see *opposite*). In life, size is often a good guide when separating this species from the smaller *C. verralli*, but size cannot be judged in photos.

Similar species: Other *Chrysotoxum* (see table on *page 104*).

Observation tips: Although it does occur inland, mainly on chalk downland in eastern England, the majority of records are from the coast of south-west England and south Wales. It has a longer flight-period than *C. cautum* and is therefore the more likely of these species to be found into August (when *C. verralli* is also flying).

J F M A M J J A S O N D

LC LC Chrysotoxum verralli

GB: Frequent =
Ir: Not recorded

Wing length: 8·25–10·5 mm

Identification: Generally much smaller than *Chrysotoxum cautum* (p. 104) and *C. elegans*, but tricky to identify. The yellow markings on the abdomen are more extensive than those of *C. elegans* so that the black front edges are narrower and more parallel-sided, broadening only at the outer corners (see *opposite*). Separation from the extremely rare *C. octomaculatum* (from which it was split in the mid-20th century) is difficult, and comparison with known specimens is required.

Similar species: Other *Chrysotoxum* (see table on *page 104*).

Observation tips: Occurs mainly in south-east England but also on the north-west coast from Liverpool to Morecambe Bay. A midsummer species which is most frequently found in hot, open grasslands or heathy rides where it will visit flowers such as Wild Parsnip and Marjoram.

J F M A M J J A S O N D

Chrysotoxum species not otherwise covered:

C. octomaculatum EN BAP		Extremely rare. The few records are from the fringes of heathlands in Surrey and Dorset.
C. vernale EN		Very rare species recorded from the coast of Hampshire and Dorset, especially the latter.

SYRPHINI: *Chrysotoxum*

black bands thicker and widening towards margin

♀ *Chrysotoxum elegans* × **6**

black bands narrower and more parallel-sided

♀ *Chrysotoxum verralli* × **6**

SYRPHINI: *DOROS* | *XANTHOGRAMMA*

from **Doros** 1 British and Irish species (illustrated)
2b
p. 98   A large, narrow-waisted hoverfly with a chocolate-coloured band along the front edge of the wing. Very little is known of its larval biology although an association with aphids tended by the black ant *Lasius fuliginosus* is suspected.

LC ## *Doros profuges*   UKBAP Priority Species
NT
GB: Scarce =

Wing length: 11·25–13·25 mm

Ir: Rare

Identification: Unmistakable due to its large size and distinctive wing markings.

Similar species: This species cannot be confused with any other British and Irish hoverfly (but might be mistaken for the genus *Ceriana* elsewhere in Europe). Confusion with thick-headed flies of the genus *Conops* (family Conopidae) is also likely.

Observation tips: Very difficult to find, even at its few well-known sites, possibly because the flight-period is very short. Most frequently noted from scrubby chalk downland and limestone grassland. Although records are mainly from southern England, one of the most consistent sites is a locality in Silverdale, Lancashire. It was also caught on the island of Mull in 1991 and was reported from a boat just off the coast of Brighton in 2017. To date there have been just six records from Ireland.

J F M A M J J A S O N D

from **Xanthogramma** 3 British and Irish species (3 illustrated)
2a
p. 98   Striking black-and-yellow hoverflies with a yellow face, a yellow stripe running along the edge of the top surface of the thorax, and yellow markings on the sides of the thorax. Only likely to be confused with *Chrysotoxum* (pp. 102–107) but lack the long, forward-projecting antennae of that genus. The larvae are believed to live in the nests of yellow or black ants of the genus *Lasius*.

LC ## *Xanthogramma citrofasciatum*  GB: Frequent =
LC
Ir: Rare

Wing length: 6·5–10·25 mm

Identification: The sharply defined, narrow, triangular marking on abdomen segment T2 together with the lemon-yellow markings and legs that are mostly orange makes this species quite easy to recognise.

Similar species: Other *Xanthogramma* species, but those have yellow legs (with black markings on the hind legs).

Observation tips: Favours short grassland with abundant Yellow Meadow Ant *Lasius flavus* nests on south-facing slopes and in sheltered glades among scrub. Locations include acid grasslands, some coastal grazing marshes and chalk grasslands. A scarce species with a predominantly southern distribution, reaching as far north as the Lake District. To date there have been just 15 records from Ireland.

J F M A M J J A S O N D

SYRPHINI: *Doros* | *Xanthogramma*

♂ *Doros profuges* × 5

♂ *Xanthogramma citrofasciatum* × 7

narrow yellow markings on abdomen segment T2

Similar species: *Xanthogramma pedissequum* and *X. stackelbergi* often require careful examination of several features for reliable identification. The most useful confirmatory character seems to be the degree to which the wing shade extends beyond veins R_{2+3}. In addition, the shape of the frons markings in the female and of the markings on abdomen segment T2 of the male are useful confirmatory features (see comparison *opposite*). *X. citrofasciatum (p. 108)* differs from both in its mostly orange legs.

Xanthogramma pedissequum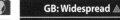

Wing length: 7·25–9·75 mm

Identification: A very distinctive black-and-yellow hoverfly that is virtually hairless. The hind femora are yellow with black apices and the hind tibiae are black. The markings on abdomen segment T2 are broad. The wing shade extends well below R_{2+3}. In females the dark band on the frons is relatively broad. **Individuals with a single large yellow marking on the side of the thorax can be assigned to *X. pedissequum* but more colourful variants require a detailed assessment of other features.**

Observation tips: Mainly a grassland species. It occurs also in woodland rides (but see *X. stackelbergi* which appears to be a woodland edge species). Mainly southern, where it is widespread, but there are a few scattered records into central Scotland.

J F M A M J J A S O N D

WING shading extends below R_{2+3}.

Xanthogramma stackelbergi

Wing length: 7·25–9·75 mm

Identification: A very brightly marked hoverfly with orange-yellow abdominal spots and very extensive yellow markings on the sides of the thorax below the wings. It is relatively hairless and has brightly marked yellow-and-black legs. **The presence of several yellow markings on the thoracic pleurae should not be considered to be conclusive proof of identity.**

Observation tips: A recent addition to the British list (2012). It appears to be a more southern species than *X. pedissequum* and is perhaps more frequently found along woodland rides and edges.

WING shading does not reach R_{2+3}.

J F M A M J J A S O N D

SYRPHINI: *Xanthogramma*

♀ | FRONS band much broader than in *X. stacklebergi*

♂ | ABDOMEN T2 marking broad

♀ | FRONS band narrower than in *X. pedissequum*

♂ | ABDOMEN T2 marking hook-shaped

♀ *Xanthogramma pedissequum* × 5

♀ *Xanthogramma stackelbergi* × 5

Sphaerophoria

from 2b
p. 98

11 British and Irish species (3 illustrated)

Small, elongate hoverflies with yellow bands or spots on the abdomen, a yellow face, yellow scutellum and prominent yellow markings on the side of the thorax. Males have a large genital capsule which forms a bulge under the end of the abdomen. **Identification of most species is only possible using male genitalia; in almost all cases females cannot, as yet, be identified to species reliably.** The larvae feed on ground-layer aphids.

LC *Sphaerophoria interrupta*

GB: Widespread =
Ir: Frequent

Wing length: 4·75–6·5 mm

Identification: Included here to represent the six species with paired spots on the abdominal segments. The male genitalia have a distinct thumb-like process. Although most female *Sphaerophoria* cannot be identified reliably, those of *S. interrupta* are distinctive in having abdominal spots and a black band down the face.

Similar species: Most other spotted *Sphaerophoria*. Note: *S. scripta* (p. 114) that develop in cold conditions can be relatively dark and may have spots rather than bands on the abdominal segments.

Observation tips: A common grassland species that often occurs in damper locations. It can be found with other *Sphaerophoria* species, most notably *S. fatarum* (N/I) and *S. philanthus* (N/I). Its Irish distribution is rather patchy but it is possibly under-recorded.

J F M A M J J A S O N D

Sphaerophoria species not otherwise covered:

ABDOMEN: **paired spots**	
S. bankowskae DD	Three records: from Essex, Northamptonshire and the Scottish Highlands.
S. fatarum	One of the commonest species; mainly northern on open heathland and moorland, but with scattered records farther south.
S. philanthus	Mainly northern on dry heaths and moorland; also heathy woodland rides.
S. potentillae VU	Known only from a few Culm grassland sites.
S. virgata NS	A scarce species of heathland and moorland with most records from central Scotland and the heaths of southern England.
ABDOMEN: **distinctive banded pattern; slightly waisted**	
S. loewi NT	A rare, mainly coastal species with the majority of records from brackish reedbeds; known only from four pre-2000 10 km squares in Ireland.
ABDOMEN: **usually banded**	
S. batava	Widely scattered records from heathland and open rides in woodland; known only from just five pre-2000 10 km squares in Ireland.
S. taeniata	Found mainly in southern England, generally in rich, unimproved grassland.
UNCERTAIN TAXON	
S. 'species B'	A single specimen from Blean Woods, Kent, that is either a separate taxon or a variation of another species.

SYRPHINI: *Sphaerophoria*

♀ *Sphaerophoria* sp. × 8

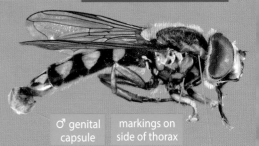
♂ *Sphaerophoria interrupta* × 8

♂ genital capsule

markings on side of thorax

Do not be confused by this stripe which is on the top edge of the thorax.

♂ genitalia

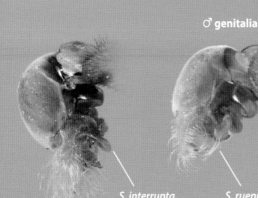
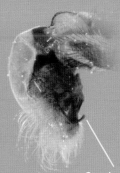

S. interrupta
'thumb'-like process

S. rueppellii
very open and hairy

S. scripta
characteristic inner process-shape

♂ *Sphaerophoria interrupta* × 10

SYRPHINI: SPHAEROPHORIA *Sphaerophoria* male genitalia compared: *p. 113*

LC *Sphaerophoria rueppellii*

LC

GB: Frequent ▽
Ir: Rare

Wing length: 4·25–6·5 mm

Identification: Males and females of this species are reasonably straightforward to identify if examined carefully. The combination of an interrupted thoracic stripe, a slightly waisted abdomen and yellow antennae is fairly distinctive in both sexes.

Similar species: *Sphaerophoria loewi* (N/I) looks very similar but has black antennae. Separation from other species with a complete thoracic stripe can be problematic when this stripe is faint: numerous examples of putative *S. rueppellii* have proven to be one of the other spotted species.

Observation tips: Often found in ruderal situations where plants such as Scentless Mayweed, Fat-hen and knotweeds are abundant, but may also be found in association with Hairy Willowherb. In Britain it is mainly a southern and coastal species that is particularly abundant around the Thames Estuary; inland records are less frequent, but not unusual. To date there have been just two records from Ireland, both from the south-east.

J F M A M J J A S O N D

> **Note:** Two species of *Sphaerophoria* (*S. rueppellii* and *S. loewi*) have an interrupted yellow stripe along the sides of the top of the thorax. In all other *Sphaerophoria* the stripe is uninterrupted, even if somewhat darkened and obscured (see *opposite*).

LC *Sphaerophoria scripta*

LC

GB: Widespread ▲
Ir: Frequent

Wing length: 5–7 mm

Identification: Males are reasonably distinctive because the abdomen is longer than the wings and protrudes well beyond the wing-tips when the wings are closed. The markings on the abdomen are usually broad yellow bands. However, the coloration is strongly influenced by the temperature at which larvae developed. Many early spring individuals are dark (see inset *opposite*) and may a have spotted, rather than banded, abdomen when development has taken place in colder environments (especially in the spring).

Similar species: *Sphaerophoria scripta* can have a spotted abdomen, so confusion with *S. interrupta*-type species is possible. Confusion with shorter-bodied banded species (mainly *S. batava* and *S. taeneata*) is also possible – in those species the abdomen length does not exceed the length of the wings when at rest. Longer black-and-yellow male *Platycheirus* (*pp. 75–93*) may also be mistaken for *S. scripta* but lack the extensive yellow markings on the thoracic pleurae.

J F M A M J J A S O N D

Observation tips: A partial migrant which probably does not have a resident population in the more northerly parts of Britain. It is the commonest *Sphaerophoria* and is widespread in grasslands in England and Wales. It is also widespread in southern Ireland but seemingly absent from much of the north, possibly reflecting the degree to which migratory specimens disperse.

SYRPHINI: *Sphaerophoria*

thoracic stripes

S. rueppellii & *S. loewi*
stripe interrupted

all other *Sphaerophoria* species
stripe uninterrupted

♂ *Sphaerophoria rueppellii* × **10**

♂ *Sphaerophoria scripta* × **10**

S. scripta ♂ (× 5) with spots rather than banded markings

SYRPHINI: *ERIOZONA* | *LEUCOZONA*   See also *Identifying wasp and bee mimics pp. 62–65*

from 3a
p. 98
Eriozona
1 British and Irish species (illustrated)
A bumblebee mimic. The larvae feed on tree-dwelling aphids.

LC
LC
Eriozona syrphoides

Nationally Scarce
GB: Frequent ▽
Ir: Local

Wing length: 12 mm

Identification: A distinctive bumblebee mimic due to its wholly yellow face (unique among British and Irish bumblebee mimics) and yellow scutellum.

Similar species: Potentially confused with other bumblebee mimics, especially *Cheilosia illustrata* (*p. 176*), which has a black face and scutellum; female *Eristalis intricaria* (*p. 228*), which has a loop in vein R4+5; and *Leucozona lucorum*, which has a partially darkened face.

Observation tips: Mainly a species of conifer plantations, but increasingly reported from south east England where larvae have been found feeding on deciduous tree aphids. Usually found in mid- to late summer in open, sunny rides, visiting flowers such as Hogweed, Wild Angelica, scabiouses and Heather. It is a relatively recent colonist (first found in Wales in the 1960s) but is now widely distributed, although rarely common. To date there have been just 12 records from Ireland.

J F M A M J J A S O N D

from 3b
p. 98
Leucozona
3 British and Irish species (all illustrated)
Species in this genus differ in colour and markings from the majority of other hoverflies, the blue-grey markings of female *L. glaucia* being particularly unusual. They are all woodland species and can be found together on Hogweed flowers. The larvae feed mainly on ground-layer aphids, although *L. lucorum* larvae have also been found feeding on arboreal aphids.

LC
LC
Leucozona lucorum

GB: Widespread =
Ir: Widespread

Wing length: 7·75–10·0 mm

Identification: Broad creamy markings on abdomen segment T2 and distinct wing clouds. Males are darker than females.

Similar species: Most likely to be confused with *Cheilosia illustrata* (*p. 176*), but that species has a black, rather than yellow, scutellum and the pale abdominal areas are due to white hair, not markings; *Volucella pellucens* (*p. 272*) [re-entrant outer cross-vein] is also frequently mis-reported as this species. May also be confused with *Eriozona syrphoides* [T2 no broad pale markings] and *Eristalis intricaria* (*p. 228*) [black scutellum and loop in wing vein R4+5].

Observation tips: Widespread along woodland rides, edges and hedgerows, but there is evidence that it is disappearing from the London suburbs and from south-east England. Visits flowers (such as Greater Stitchwort and Garlic Mustard) in dappled sunshine. Although mainly a spring species, it sometimes has a much less abundant second generation in midsummer, especially in Wales and more northerly regions.

J F M A M J J A S O N D

Identification features of *Leucozona*: *p. 118*

SYRPHINI: *Eriozona* | *Leucozona*

♂ *Eriozona syrphoides* × **4** ♀

♂ *Leucozona lucorum* × **7**

Identification features of *Leucozona*

SPECIES	KEY FEATURES
L. lucorum (p. 116)	ABDOMEN Distinctive, with broad, creamy markings on T2 WING distinct clouds
L. glaucia (below)	♂ similar to *L. laternaria* but scutellum and front legs **yellow** ♀ ABDOMEN Unmistakable blue-grey markings, and a yellow scutellum and front legs
L. laternaria (below)	similar to *L. glaucia* but scutellum and front legs **darkened**

Leucozona glaucia

GB: Widespread ▽
Ir: Widespread

Wing length: 8·0–11·25 mm

Identification: The blue-grey abdominal markings of females are unmistakable. Males have a yellow scutellum and the front legs are mainly yellow.

Similar species: This species can really only be confused with *Leucozona laternaria*, in which the scutellum and front legs are dark.

Observation tips: Mainly a western and northern species, which is also well-represented across Ireland. Once widespread in the wooded counties of south-east England, it has now largely disappeared from these areas. The reasons for this change are thought to be climate-related. It is still abundant in the north, where it can be one of the commonest hoverflies in midsummer on Hogweed and Wild Angelica in woodland rides.

J F M A M J J A S O N D

Leucozona laternaria

GB: Widespread =
Ir: Frequent

Wing length: 7–10 mm

Identification: Males are sometimes a lot darker than females and can often be poorly marked. Both are a lot darker in appearance than the other two species in this genus. In females, the combination of strong white markings on T2 and bluish-grey markings on T3 & T4 mean that it is normally readily recognised.

Similar species: Apart from *Leucozona glaucia* (both sexes), which has a yellow scutellum and front legs, some very dark males may be confused with *Melangyna quadrimaculata* (p. 158), which has a black face (pale in *L. laternaria*).

Observation tips: A woodland species which occurs most frequently on umbellifers, especially Hogweed and Wild Angelica. Its larvae are particularly associated with aphids on umbellifers. It is slightly more widespread in eastern England than *L. glaucia*, but is still less abundant here than it is farther west. In common with parts of Britain, Irish data suggest that this species is less abundant than *L. glaucia* but has a similar distribution.

J F M A M J J A S O N D

SYRPHINI: *Leucozona*

♂ *Leucozona glaucia* ×5 ♀

♀ *Leucozona laternaria* ×5

SYRPHINI: *DASYSYRPHUS*

from 3c
p. 98

Dasysyrphus 8 British and Irish species (4 illustrated)

Yellow-and-black hoverflies with hairy eyes, normally with a central black stripe on the face and a **distinct, elongate black stigma** (the marking along the front edge of the wing). Several of the species have distinctive markings that make them straightforward to identify. However, there is a group of species which includes *D. venustus* (see *page 122*) that is much more challenging. The larvae are camouflaged with bark-like colour patterns (see *page 17*). They spend the day resting on twigs and branches and the night feeding on aphids on both coniferous and deciduous trees.

LC *Dasysyrphus albostriatus*

LC

GB: Widespread ▽
Ir: Frequent

Wing length: 6·25–9·0 mm

Identification: The obliquely angled bars on abdomen segments T2–T4 and the pair of grey stripes on the thorax make this species straightforward to identify.

Similar species: *Didea* (*p. 124*), arguably, have similar downward-pointing markings on abdomen segments T2 and T3, but the markings are broader; the overall form of *Didea* is broader and more robust, and the wing venation is different.

Observation tips: There are two distinct generations of this widespread woodland species, one in the spring and another in late summer and early autumn. Both sexes visit flowers and bask on sunlit leaves along woodland rides. In Britain, this is a widespread species south of the Scottish border but farther north it is mainly confined to the lowlands and to coastal situations. In Ireland it is seemingly most frequently recorded in the south, with limited representation in the north-west.

J F M A M J J A S O N D

LC *Dasysyrphus tricinctus*

LC

GB: Widespread ▽
Ir: Local

Wing length: 7·25–10·25 mm

Identification: The abdominal markings consist of strong bars on abdomen segment T3 and narrow bars on segment T4. This pattern is relatively distinct and should make it readily identifiable.

Similar species: Confusion with *Chrysotoxum bicinctum* (*p. 102*) might occur under some circumstances, but that species has extensive darkening along the leading edge of the wing and yellow markings on the side of the thorax which should eliminate any confusion.

Observation tips: A woodland species which is most frequently encountered in coniferous woodlands and on heathlands. Adults are flower visitors, favouring open yellow flowers such as buttercups and dandelions. Although widespread in Britain, it is seemingly absent from much of Ireland, with a few localised areas of occurrence.

J F M A M J J A S O N D

SYRPHINI: *Dasysyrphus*

Dasysyrphus have a distinctive elongate black stigma.

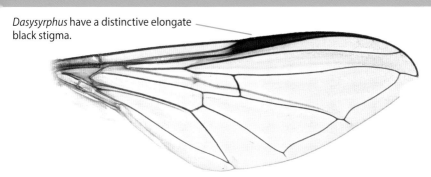

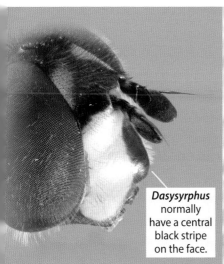

Dasysyrphus normally have a central black stripe on the face.

♂ *Dasysyrphus albostriatus* × 6

grey stripes on thorax

♂ *Dasysyrphus tricinctus* × 6

LC *Dasysyrphus venustus* sensu lato

GB: Widespread ▽
Ir: Local

Wing length: 6·25–10·0 mm

Identification: The narrow, hook-shaped bars on abdomen segments T3 and T4 are characteristic of a number of very similar species (see *below*). In the *D. venustus* complex these bars reach the edge of the abdomen. In addition, the black markings of abdomen underside segment S2 (see *opposite*), although variable, are typically deeper in the middle than at the edges.

Similar species: Other members of the *Dasysyrphus venustus* complex (see table *below*); *D. pinastri* and *D. pauxillus*, in which the yellow bars on T3 and T4 do not reach the abdomen edge and are separated by a narrow black strip (see inset *opposite*). Confusion is possible with some *Eupeodes*, often with *E. luniger* (*p. 147*), but in *Eupeodes* the stigma is pale (strongly darkened in *Dasysyrphus*).

D. venustus has been split into two species, with a new segregate (*D. neovenustus*) recognised on the basis of a range of characters, the shape of the black markings of abdomen underside segment S2 (see *opposite, top*) seemingly being the most reliable.

J F M A M J J A S O N D

Observation tips: A species of both deciduous and coniferous woodlands across lowland Britain that basks on sunny leaves and visits flowers such as buttercups and *Acer* species. Both *D. venustus* and *D. neovenustus* have been recorded in Ireland, although there are very few records.

An explanation of the *Dasysyrphus venustus* complex

D. venustus	Widespread in Britain. Check the shape of black marking on sternite 2 (see *opposite*).
D. neovenustus	Seemingly more frequent in northern Britain. Check the shape of black marking on sternite 2 (see *opposite*).
D. friuliensis	Very few recent records. Difficult to separate from *D. venustus* but worth checking larger specimens with deeply indented abdominal markings (specimen needed for comparative assessment).
D. hilaris	Very few recent records but mainly a northern species. Difficult to separate from *D. venustus* but check for completely yellow face.
Note These are a complex of species around *D. venustus*, which was revised in 2013. Critical characters are somewhat variable and not all specimens readily run to one of the segregates. Both *D. hilaris* and *D. neovenustus* are known only from pre-2000 records in Ireland. **The difficulty of separating *D. venustus* from *D. hilaris* and *D. neovenustus* means that individuals should be recorded as *D. venustus* sensu lato (which means in the broad sense) unless a specimen has been retained and examined microscopically. Reliable identification from photographs is generally not possible without a view of the face and the underside of the abdomen.**	

Other *Dasysyrphus* species not otherwise covered:

D. pinastri/ *D. pauxillus*	The *D. pinastri* complex occurs throughout Britain in conifer plantations but is most frequent in northern England and Scotland. There are very few records of *D. pauxillus*.

SYRPHINI: *Dasysyrphus*

Dasysyrphus venustus × **6**

abdomen undersides

D. venustus

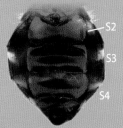

S2
S3
S4

black on S2 deeper in middle than at the edges (but very variable)

D. neovenustus

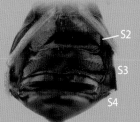

S2
S3
S4

D. neovenustus black on S2 forms a parallel-sided band.

D. venustus complex – yellow bars on T3 & T4 **reach** the abdomen edge

D. pinastri/D. pauxillus – yellow bars on T3 & T4 **do not reach** the abdomen edge

Dasysyrphus pinastri ♀ × **6**

SYRPHINI: *DIDEA*

from
4a
p. 99

Didea

3 British and Irish species (2 illustrated)

This is a highly distinctive genus, for two reasons. Firstly, **there is a marked loop in wing vein R$_{4+5}$** (see *opposite*), although this is not as pronounced as in the Eristalini (see *p. 218*); this is an unusual feature among the yellow-and-black hoverflies. Secondly, the abdominal markings are obliquely angled – distinctive and different from almost all other species. The larvae feed mainly on aphids on conifers. Only one species, *Didea fasciata*, is at all abundant.

LC *Didea fasciata*
LC

GB: Frequent ▽
Ir: Frequent

Wing length: 8·25–11·0 mm

Identification: Yellow-and-black hoverfly with distinctive obliquely angled abdominal markings, yellow haltere clubs and a largely yellow scutellum. A green-blue variant is known, which might be confused with *Didea alneti* (N/I).

Observation tips: A woodland hoverfly which has two marked peaks of emergence: a small one in the spring and another, much larger, one in late summer. Males are strongly territorial and can be seen defending sunlit leaves. It occurs throughout mainland Britain but is predominantly a southern species. In Ireland, it is patchily distributed and appears to favour more coastal locations.

Similar species: The two resident *Didea* species are readily separated by haltere and scutellum coloration. *Dasysyrphus albostriatus* (*p. 120*) has similar (although narrower) abdominal markings but is more delicate in build, with wings that have a long, dark stigma and lack a marked loop in R$_{4+5}$.

J F M A M J J A S O N D

LC *Didea intermedia*
LC

Nationally Scarce
GB: Scarce =
Ir: Not recorded

Wing length: 7–10 mm

Identification: As *Didea fasciata*, a yellow-and-black hoverfly with distinctive oblique abdominal markings, but with black haltere clubs and a yellow scutellum that is extensively darkened along the rim.

Observation tips: Adults fly mainly from June to August and seem not to stray far from conifers. They can be found on yellow composites on heathland with conifers in southern England, and in conifer plantations in the north and west. In Scotland, it can be quite abundant along rides in coniferous woodlands and plantations where it frequently visits the flowers of Heath Bedstraw.

J F M A M J J A S O N D

Didea species not otherwise covered:

D. alneti	A very rare vagrant with abdominal markings that are usually greenish or turquoise, rather than yellow. The last known records were from Northumberland in 1989.

SYRPHINI: *Didea*

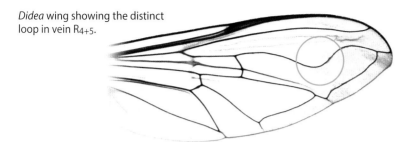

Didea wing showing the distinct loop in vein R$_{4+5}$.

***Didea* halteres**
The colour of the haltere club separates the two *Didea* species. This character is often hidden under the wings and will not necessarily be shown by a photograph.

♂ *Didea fasciata* × **6**

D. fasciata
yellow

♀ *Didea intermedia* × **5**

D. intermedia
black

SYRPHINI: *SCAEVA*

from 5
p. 99
Scaeva

5 British and Irish species (2 illustrated)

Black-and-white or yellow-marked hoverflies with hairy eyes, and in which the area between the eyes bulges strongly (more so than in any other British hoverfly). Whilst *S. selenitica* may be resident, the commonest species, *S. pyrastri*, is a migrant. The other three species (*S. albomaculata*, *S. mecogramma* and *S. dignota*) are rare vagrants. The larvae feed on aphids.

Scaeva selenitica

GB: Frequent =
Ir: Local

Wing length: 10·5–12·0 mm

Identification: A relatively large hoverfly with relatively deeply curved, comma-shaped, yellow abdominal markings that have their rear edge parallel to the rear margin of the segment. However, it is quite variable, making positive identification tricky in a few cases.

Similar species: *Scaeva dignota* (N/I) males have a narrower head; in females the abdominal markings reach the tergite margin. *Eupeodes lundbecki* (see p. 146) (N/I) has a strongly dipped vein R_{4+5} but the eyes are bare and the frons is not significantly inflated. Some newly emerged *S. pyrasti* can be yellowish and cause confusion.

Observation tips: Most frequent in the vicinity of coniferous woodland or on heathland with conifers, suggesting that there is a resident population – possibly augmented by migrants. The larvae have been found feeding on aphids on pines. Widely distributed in Britain but in Ireland it appears to be predominantly southern.

J F M A M J J A S O N D

Scaeva pyrastri

GB: Widespread =
Ir: Widespread

Wing length: 9·25–12·5 mm

Identification: A relatively large hoverfly with broad white abdominal markings, the rear edges of which lie obliquely to the rear margin of the segment. In combination with the swollen frons, the markings make it distinctive. Entirely dark individuals can occur, but are usually recognisable by the swollen yellow frons.

Similar species: Other *Scaeva* with separated markings, especially when viewed from awkward angles or if immature. Most likely to be confused with *S. selenitica* and *S. dignota*, both of which have yellow markings that are more comma-shaped.

Observation tips: Migratory, arriving in Britain and Ireland annually in highly variable numbers: if good numbers arrive, then localised breeding may occur. Can turn up almost anywhere, but scarcer in the uplands of Britain and in north-west Ireland

J F M A M J J A S O N D

Scaeva species not otherwise covered:

S. albomaculata	Vagrant from southern Europe with two known records from the south coast of England in 1938 & 1949.
S. dignota	An addition to the British list in 2013 which bears a very close resemblance to *S. selenitica*. Easily overlooked.
S. mecogramma	A single record from Aniston, Lothian in 1905. Likely an accidental import.

Identifying hoverflies with yellow-and-black banded abdomen patterns – pitfalls and process

Species with yellow bands on the abdomen cause problems for most novices as well as for many more experienced photographic recorders. This is a frequently recurring problem on iRecord and on the UK Hoverflies Facebook group (see *page 338*).

Recognition based on band shape and colour is highly problematic and unfortunately the most useful character, the presence or absence of squamal hairs, cannot be seen on live animals or in photographs.

There is no escaping the basic challenge of how to learn to recognise *Syrphus*. The genus (apart from *S. nitidifrons*) has a certain jizz but that is not much help to a novice. The most reliable way of telling *Syrphus* from other yellow-banded Syrphini is the presence of upstanding hairs on the upperside of the squama (a membrane located where the hind margin of the wing meets the thorax).

squama

hairs on upper surface = *Syrphus* (see note *below*)	bare upper surface = other banded Syrphini
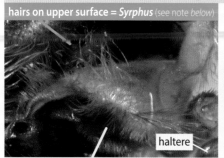 haltere	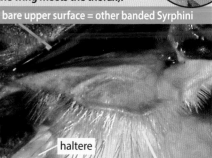 haltere

NOTE: Ignore the hairs on the margin of the squamae.

Unless specimens are retained and checked under a microscope it is almost impossible to see the hairs on the squamae. The following guide should help to separate these banded genera without the need to check the squamae but it is not foolproof and still relies upon a degree of experience.

As always, if specimens are not being retained it is essential to try and get photographs from three angles – face on, top down and side on.

1 Face colour		2 Shape of wing vein R_{4+5}	
Beware – *Parasyrphus* are very difficult to separate from *Syrphus* in photos unless a face shot has been secured. Hind leg colour may act as a clue but it is difficult to be conclusive.		strongly dipped	straight
central darkened stripe	completely yellow	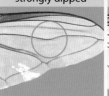	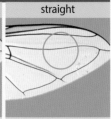
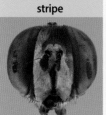		▶ *Megasyrphus* (*p. 131*)	*Syrphus, Parasyrphus, Eupeodes* and *Epistrophe*
▶ most *Parasyrphus* (*p. 142*)	*Parasyrphus nigritarsis, Syrphus, Megasyrphus, Epistrophe* and *Eupeodes*	NOTE: The dip is very distinctive but is also present in *Didea* (*p. 124*) which might be misinterpreted as a banded species.	

SYRPHINI

3 Colour of hairs on segment margins

completely black (typically very short and uniform length)

partly yellow (variable in length; usually long)

typically short and uniform

usually long and variable

▶ *Eupeodes* (p. 146)

Syrphus, *Parasyrphus*, and *Epistrophe*

Beware – *E. latifasciatus* (p. 148) is often confused with *Syrphus* but has a distinctive black-and-yellow frons which lacks any dusting (see *below*).

4 Abdomen segment 2 markings

yellow area reaches the margin over a much broader extent than in *Syrphus* or *Parasyrphus*

yellow area 'golf club head'-shaped (lower area of black extends upwards at the margin, narrowing the extent of yellow that reaches the margin)

▶ *Epistrophe* (p. 136)

Syrphus (p. 132) and *Parasyrphus* (p. 142)

NOTE: *Epistrophe* tends to have an orange hue that fades over time, especially in preserved specimens.

NOTE: Most *Parasyrphus* look so similar to *Syrphus* that they are readily overlooked and misidentified; they are most likely to be picked out by checking the face for a black stripe.

Beware – *P. nigritarsis* (p. 144) has a yellow face but has a more orange hue to the abdominal markings.

Using the frons in females for identification

In females there can be clues on the frons that help separate *Syrphus* from some *Epistrophe* and banded *Eupeodes* as shown *below*:

SYRPHUS (p. 132)
the frons is extensively dusted

Eupeodes latifasciatus (p. 148)
the frons is approximately half black and half yellow

Epistrophe diaphana (p. 138)
the frons has no dusting and is mostly yellow

YELLOW-AND-BLACK BANDED SYRPHINI

Banded Syrphini compared – a cautionary tale

Making a determination based on band shape is fraught with difficulty as many species look very similar – especially *Syrphus*, some *Epistrophe* and banded *Parsasyrphus*. The following are images of the first band on the abdomen of representative species to illustrate their similarity and the need to use other, more detailed, features (such as the length and colour of the hairs on the segment margins that make *Eupeodes* distinctive) to improve the likelihood of an accurate identification.

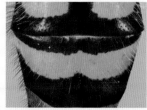

PARASYRPHUS (p. 142)

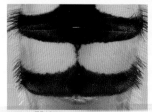

EPISTROPHE (p. 136)

SYRPHUS (p. 132)

Epistrophe grossulariae (p. 138)

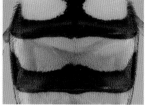

Eupeodes latifasciatus (p. 148)

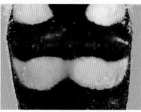

Eupeodes bucculatus (p. 146)

A tip for finding some aphid-feeding larvae

Some species of aphid-feeding larvae are seemingly easy to find if you know where to look! One of the most productive places is on gravestones beneath Sycamore trees after high winds during the late summer and autumn. At this time, both hoverfly larvae and aphids get displaced, shaken to the ground or can be found on leaves that have been blown from the trees. Both larvae and aphids tend to ascend the gravestones where a diligent observer may find hoverfly larvae feeding on the aphids. The most abundant larvae are from the genus *Syrphus*, which can only be identified by rearing them through to an adult. *Dasysyrphus* larvae, especially *D. albostriatus*, are also found infrequently, together with some *Epistrophe*. If you do not have a nearby graveyard, an alternative is to check out fence posts and rails beneath trees – you never quite know what you will find. And, if rearing larvae, do not be disappointed if you have the luck to get a parasite – remarkably little is known about parasites of hoverflies so there are opportunities to increase knowledge of this subject: do keep them and pass them on to a specialist (via the UK Hoverflies Larval Group Facebook page, see *page 338*).

Larva of *Syrphus* sp.

SYRPHINI: *Megasyrphus*

Megasyrphus

1 British and Irish species (illustrated)

This is a genus with a single species which has distinct black-and-yellow bands on the abdomen. It is likely to be overlooked as a *Syrphus* (pp. 132–135). Closer inspection reveals a distinct dip in the vein R_{4+5}, and a broad black stripe across the undersides of the abdominal segments. The larvae feed on a variety of aphids, especially those feeding on broadleaved trees such as Alder and Ash.

Megasyrphus erraticus

Wing length: 10·75–12·0 mm

Identification: A large species with a strong dip in vein R_{4+5}, a shiny black thoracic dorsum, and a yellow-and-black banded abdomen. The face is yellow with a black central stripe. In life, it tends to be more orange than yellow but this feature fades with age and in preserved specimens.

Similar species: Most likely to be confused with *Syrphus* (*pp. 132–135*), which differ in having an unmarked yellow face and an olive-coloured (and somewhat dull) thorax. Confusion with banded *Epistrophe* (*pp. 136–141*) and with *Parasyrphus nigritarsis* (*p. 144*) is also possible, but all these three genera lack the distinct dip in wing vein R_{4+5}.

Observation tips: This is primarily a northern and western species which is found mainly in coniferous plantations. It will visit a wide variety of flowers, especially yellow composites. Rides or clearings with a stream are good places to look. There are very few records from Ireland and only one since 2000.

Nationally Scarce
GB: Local ▽
Ir: Rare

J F M A M J J A S O N D

Megasyrphus erraticus × 4

dip in vein R_{4+5}

SYRPHINI: *SYRPHUS*

Yellow-and-black banded hoverflies in the genera *Epistrophe, Eupeodes, Syrphus, Parasyrphus* and *Megasyrphus* are consistently misidentified. See *pages 128–130* for a guide to their identification.

from 5
p. 99

Syrphus
5 British and Irish species (3 illustrated)

Yellow-and-black banded hoverflies with a yellow face and a dusted thorax that has a dull, bronzy-greenish appearance. The upstanding hairs on the upper surface of the squamae is diagnostic of the genus *Syrphus* among yellow-banded Syrphini (see *page 100*). *Syrphus ribesii, S. torvus* and *S. vitripennis* are among the commonest yellow-and-black banded flower-visiting hoverflies, but are difficult to separate. There are two other British species: *S. rectus* (probably not valid as a European species); and *S. nitidifrons*, a recent addition to the British list. The larvae feed on aphids. *Syrphus* are frequently misidentified as other genera – see *page 128*.

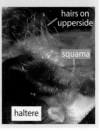

Identification features of *Syrphus*

SPECIES	MALES: Eyes touching on top of head		WING (both sexes) (see *page 134*)	FEMALES: Eyes well separated on top of head
S. ribesii (below)	EYES	HIND FEMUR yellow area at end with at least **some black hairs**	2nd basal cell entirely covered by microtrichia	HIND FEMUR **mostly yellow**, black at extreme base only
S. vitripennis (p. 134)	bare	HIND FEMUR yellow area at end **with yellow hairs**	2nd basal cell **partially** covered by microtrichia	HIND FEMUR extensively darkened, last ¼–⅓ yellow
S. torvus (p. 134)	EYES **hairy**		2nd basal cell entirely covered by microtrichia	
S. rectus	There is growing doubt that this is a valid British species putative specimens seem to be pale-legged female *S. vitripennis*			
S. nitidifrons	Most likely to be confused with *Parasyrphus punctulatus* (p. 142) as hairs on the squamae are often sparse and inconspicuous.			

LC
LC

Syrphus ribesii

GB: Widespread =
Ir: Widespread

Wing length: 7·25–11·5 mm

Identification: Yellow-and-black banded, with a yellow face, bare eyes and wing basal cells completely covered in microtrichia. The shape of the bands varies enormously between individuals, in some almost separated into spots. Although hair colour on the hind femora can be a helpful character in males (*opposite*), the key feature is the combination of bare eyes and complete microtrichial cover (requires microscopy and experience to interpret). In females, the hind femora are almost completely yellow, a feature unique to British and Irish *Syrphus* (but not so in mainland Europe).

Similar species: All banded *Syrphus*, some *Epistrophe* (pp. 136–141) (especially *E. diaphana* and *E. grossulariae*), *Megasyrphus erraticus* (p. 131) and *Parasyrphus* (pp. 142–145).

Observation tips: Occurs from early spring to autumn in the lowlands. Males often form swarms hovering in dappled light under trees and emit much of the familiar buzz of a spring woodland.

J F M A M J J A S O N D

SYRPHINI: *Syrphus*

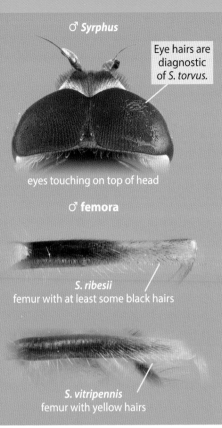
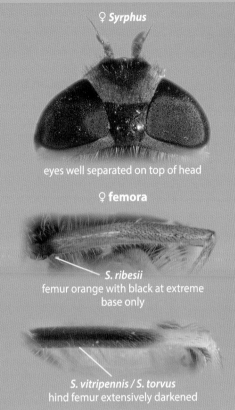

♂ *Syrphus*

Eye hairs are diagnostic of *S. torvus*.

eyes touching on top of head

♀ *Syrphus*

eyes well separated on top of head

♂ **femora**

S. ribesii
femur with at least some black hairs

S. vitripennis
femur with yellow hairs

♀ **femora**

S. ribesii
femur orange with black at extreme base only

S. vitripennis / *S. torvus*
hind femur extensively darkened

♀ *Syrphus ribesii* × 6

Syrphus torvus

GB: Widespread =
Ir: Frequent

Wing length: 8·5–11·75 mm

Identification: Yellow-and-black banded, with a yellow face. The hind femora are black-and-yellow in both sexes. The eyes of males are covered in pale hairs which can be seen through a 20× hand lens. In females, the eye hairs are tiny, sparse and very hard to see; if uncertain, check the 2nd basal cell of the wing, which in both sexes is completely covered with microtrichia (see *below*).

Observation tips: Widespread throughout lowland Britain and most frequently encountered in the spring. In Ireland, it appears to be most widespread and abundant in the south.

Similar species: All banded *Syrphus*. Some *Epistrophe*, especially *E. diaphana* and *E. grossulariae* (pp. 136–141), *Eupeodes latifasciatus* (p. 148), *Megasyrphus* (p. 131) and *Parasyrphus* (pp. 142–145) which all have bare upper surface of the squamae.

J F M A M J J A S O N D

Syrphus vitripennis

GB: Widespread =
Ir: Frequent

Wing length: 7·25–10·25 mm

Identification: Yellow-and-black banded, with a yellow face. The hind femora are black-and-yellow in both sexes. The eyes are bare and the second basal cell of the wing is partly free of microtrichia (see *below*). The shape of the yellow bands varies enormously between individuals, some being almost separated into spots. Although hair colour on the hind femora can be a helpful character in males (see *page 133*), the key feature is the combination of bare eyes and partly bare second basal cell (requires microscopy and experience to interpret).

Observation tips: Abundant in lowland Britain, but numbers fluctuate enormously due to mass immigration from continental Europe. In Ireland, it appears to be much more widespread and abundant in the south. Can be a common visitor to garden flowers in midsummer, with numbers bolstered by migrants.

J F M A M J J A S O N D

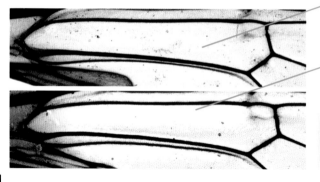

S. ribesii and *S. torvus*: entirely covered with microtrichia

S. vitripennis: partially covered with microtrichia, leaving a bare margin along the front of the cell

Wings of *S. torvus* and *S. vitripennis* showing the characteristics of the 2nd basal cell.

SYRPHINI: *Syrphus*

♂ *Syrphus torvus* ×**6**

♂ eyes obviously hairy (hairs usually indistinct in ♀)

♀ *Syrphus vitripennis* ×**6**

SYRPHINI: *EPISTROPHE*

> Yellow-and-black banded hoverflies in the genera *Epistrophe, Eupeodes, Syrphus, Parasyrphus* and *Megasyrphus* are consistently misidentified. See *pages 128–130* for a guide to their identification.

from 5 p.99

Epistrophe

7 British and Irish species (5 illustrated)

Species with yellow or orange bands and mainly yellow faces that fly mostly in the spring and early summer (except for *Epistrophe grossulariae*, which flies later). Apart from the distinctively marked (although very variable) *E. eligans* they are often confused with *Syrphus* (*pp. 132–135*) and *Parasyrphus* (*pp. 142–145*). *Epistrophe*, unlike *Syrphus*, lack upstanding hairs on the upper surface the squamae (but they do have long hairs around the margin, as in other yellow-banded Syrphines) – see *page 132*. There are several additional European species that could turn up in Britain and **the genus includes some species whose identification is more challenging**. The larvae feed on tree-dwelling aphids, with those on Sycamore especially favoured by *E. grossulariae*.

LC *Epistrophe eligans*

LC

GB: Widespread ▲
Ir: Frequent

Wing length: 6·25–9·5 mm

Identification: A shining, brassy-coloured hoverfly with yellow markings mainly on abdomen segment T2 and absent or much reduced on the other segments, but very variable.

Similar species: Unlikely to be mistaken, although some small *Eristalis* (*pp. 220–233*) may look superficially similar (but have a loop in veins R$_{4+5}$).

Observation tips: Primarily a spring species which visits a wide range of flowers. The larvae are associated with aphids on trees and shrubs, favouring Elder, Sycamore and fruit trees. Males often form swarms, competing for sunlit spots in woodland rides or around a particular bush. This is one of a small number of species that are highly responsive to warmer springs and it now emerges much earlier than it did even 20 years ago. It occurs into northern England and coastal and lowland Scotland, but is much less common farther north. In Ireland, it appears to be most widespread and abundant in the south.

J F M A M J J A S O N D

The larvae of *Epistrophe eligans* are frequently found in gardens. They feed on aphids on a variety of shrubs and herbaceous plants, so adults may be seen ovipositing close to aphid colonies. Several eggs are laid close to these colonies and the larvae eat the aphids. Late in the summer, mature (third instar) larvae seek out sheltered places such as under branches or pieces of debris lying on the ground to pass the winter. During this time they are in diapause, a form of suspended development that allows the animal to survive adverse conditions – this stage is obvious because the larvae change colour from a delicate green to orange and brown hues. Once inclement weather has passed, the larva changes into a puparium before a new adult emerges in the spring. This life-cycle is depicted *opposite*.

***Epistrophe* species not otherwise covered** (both banded, with orange antennae):

E. flava	DD	An addition to the British fauna in 2022, from a 1988 specimen collected at Dungeness, Kent, which probably represents a vagrant. Identification requires detailed microscopic examination of a range of characters and comparison with specimens of *E. nitidicollis* and *E. melanostoma*.
E. ochrostoma	DD	A single, somewhat doubtful, record from north Wales in 1990.

SYRPHINI: *Epistrophe*

Epistrophe eligans ×6

Life stages of *E. eligans*

female ovipositing

larva feeding

larva in diapause

puparia

Epistrophe diaphana

GB: Frequent =
Ir: Not recorded

Wing length: 7·25–9·75 mm

Identification: A banded *Epistrophe* with black antennae and a yellow face. Frons yellow with no dusting (male), or yellow-and-black (female). Abdominal markings almost parallel with slight central indentation and strongly upturned at the tergite margins. All femora and tibiae are yellow.

Similar species: This is one of two *Epistrophe* species with black antennae (*E. grossulariae* has straight abdominal bands downturned at the margins). Other *Epistrophe* have orange or muddy-brown antennae. The main confusion is with *Syrphus ribesii* (p. 132), although confusion with banded *Eupeodes* (p. 146) and *Parasyrphus* (p. 142) is also possible. However, *Epistrophe diaphana* is distinctive in having no dusting on the frons (although beware that female *Eupeodes latifasciatus* (p. 148) has similar frons characteristics).

J F M A M J J A S O N D

Observation tips: A woodland and scrub-edge hoverfly, often seen visiting Hogweed and other umbellifers. Although a southern species, it has become more frequent in the midlands in recent years and is slowly expanding its range northwards.

Epistrophe grossulariae

GB: Widespread ▽
Ir: Frequent

Wing length: 9·0–12·25 mm

Identification: A banded *Epistrophe* with black antennae and a yellow face. Frons dusted yellow with black above the antennae (male), or dusted yellow with a strong black central stripe (female). Abdominal markings almost parallel and downturned at the tergite margins. All femora yellow with a black base; tibiae yellow.

Similar species: This is one of two *Epistrophe* species with black antennae (*E. diaphana* has abdominal bands strongly upturned at the margins). Other *Epistrophe* have orange or muddy-brown antennae. The main confusion is with *Syrphus ribesii* (p. 132), although confusion with banded *Eupeodes* (p. 146) and *Parasyrphus* (p. 142) is also possible. The almost parallel markings of *Epistrophe grossulariae* and its completely black antennae should set it apart from other similar species because the bands of those species are more moustache-shaped.

J F M A M J J A S O N D

Observation tips: A woodland edge species which appears to be more frequent in the damper environments of northern and western Britain. It is seemingly absent from much of southern Ireland but has a patchy and local distribution in the northern half.

Epistrophe grossulariae regularly visits Devil's-bit Scabious.

SYRPHINI: *Epistrophe*

margin of band upturned at edge

♀ *Epistrophe diaphana* × **6**

♀ *Epistrophe grossulariae* × **6**

margin of band downturned at edge

Epistrophe melanostoma

LC

Wing length: 9·0–12·25 mm

Identification: A banded *Epistrophe* with orange or muddy-brown antennae and a yellow face. Frons dusted yellow, with a dark wedge above the antennae (male), or a dark almost trident-like marking (female). Abdominal markings almost upturned in the middle and at the tergite margins. Scutellum with only yellow hairs. All femora yellow with a black base; tibiae yellow.

Similar species: *Epistrophe nitidicollis* (which has a yellow mouth margin), *E. flava* (N/I) and *E. ochrostoma* (N/I) (see p. 136). Views from several angles may be needed to determine the shape and coloration of frons markings and the presence of a darkened mouth margin from photographs.

Observation tips: A regular visitor to flowers such as Garlic Mustard and a frequent leaf-basker. Its range is slowly expanding to the north and west.

♀ *Epistrophe melanostoma* and *E. nitidicollis* frons patterns   J F M A M J J A S O N D

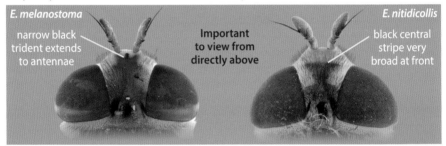

E. melanostoma — narrow black trident extends to antennae

Important to view from directly above

E. nitidicollis — black central stripe very broad at front

Epistrophe nitidicollis

LC

Wing length: 8·0–11·25 mm

Identification: A banded *Epistrophe* with orange or muddy-brown antennae and a yellow face. Frons dusted yellow, with a dark wedge above the antennae (male), or a broad dark wedge-shaped marking that is separated from the antennae by a yellow section (female) (difficult to interpret from some angles due to dusting). Black bands slightly upturned at the tergite margins. Scutellum with a mixture of yellow and black hairs: 50% or more black hairs. All femora yellow with a black base; tibiae yellow.

Similar species: *Epistrophe melanostoma* (which has a black mouth margin and all-yellow scutellar hairs), *E. flava* (N/I) and *E. ochrostoma* (N/I) (see p. 136). May be overlooked as a *Syrphus* (pp. 132–135), but in life it usually has a more orange hue. *Syrphus* species are frequently misidentified as *E. nitidicollis*.

Observation tips: See *Epistrophe melanostoma*. *E. nitidicollis* is seemingly absent from much of Ireland, although there are a few localised areas of occurrence.

J F M A M J J A S O N D

SYRPHINI: *Epistrophe*

Epistrophe melanostoma × 6

♀ ♂

Epistrophe scutella

♀ *Epistrophe nitidicollis* × 6

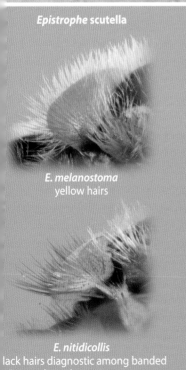

E. melanostoma
yellow hairs

E. nitidicollis
lack hairs diagnostic among banded
Epistrophe with orange antennae

Parasyrphus

from 5 p.99

6 British and Irish species (3 illustrated)

Most are black-and-yellow banded hoverflies (*P. punctulatus* has spots), the majority of which have mainly dark hind legs. The squamae lack hairs on the upper surface (see *page 128*). Most likely to be confused with *Syrphus* (*pp. 132–135*) but the presence of a dark stripe on the face and almost completely black hind femora are helpful pointers away from that genus (which is best told by the hairy upper surface of the squamae). Recognition from photos is especially challenging. Most species are associated with coniferous woodlands and many are difficult to identify because the degree of yellow marking on the 'knees' of the hind legs is individually variable both in extent and intensity. The larvae feed on aphids on trees, especially conifer aphids.

Parasyrphus punctulatus

GB: Widespread =
Ir: Frequent

Wing length: 5·5–7·75 mm

Identification: Both sexes have spots on the abdomen that reach the tergite margin: in males, these markings resemble the head of a golf-club, but in females the shape is less obvious. The face has a central dark area (can be extensively black) and the legs are mainly dark (females have yellow front tibiae). Melanistic females occur periodically in highly localised populations, but such individuals have a clearly yellow scutellum and a somewhat bronzy thorax. The scutellar hairs are black.

Similar species: Most likely to be confused with female *Melangyna lasiophthalma* (*p. 154*), which has pale hairs on the scutellum. *Syrphus nitidifrons* (N/I – but see *page 132*) is also very similar and easily overlooked – it has upright hairs on the squamae (very difficult to see), a more extensively darkened frons, paler legs and a completely yellow face. Some banded *Syrphus* (*pp. 132–135*) females have the abdominal bands almost completely interrupted but, again, have a completely yellow face. *Eupeodes* (*pp. 146–151*) species with a spotted abdomen may cause confusion, but have short, uniform hairs on the tergite margins and the markings have a less pronounced a golf-club shape.

J F M A M J J A S O N D

Observation tips: Commonest in the early spring when it visits the blossom of trees such as willows and Bird Cherry (often flying high up), but can be found through to midsummer in smaller numbers. Most common in woodland, particularly coniferous woodlands. It occurs throughout Britain but in Ireland it is mainly a southern species.

Parasyrphus species not otherwise covered:

P. annulatus	A comparatively small *Parasyrphus* with the hind femur yellow at the base (see *page 145*). Widespread but very local, usually in conifer woods; known from just four pre-2000 10 km squares in Ireland.
P. lineola	This species has black antennae and entirely black hind legs (some can be a little yellow around the 'knee'). Widespread and not uncommon in conifer woods; known from just six pre-2000 10 km squares in Ireland. It can be overlooked among *Meliscaeva cinctella* (*p. 160*).
P. malinellus	Rather similar to *P. lineola*, but usually with a more extensively yellow 'knee' (see *page 145*). The thorax is relatively shiny (dull in others species). Widespread in conifer woods, but most abundant in Scotland.

SYRPHINI: *Parasyrphus*

Parasyrphus abdomens

P. punctulatus — all other *Parasyrphus*

golf-club-shaped spots on segments T3 and T4

moustache-shaped bands on segments T3 and T4

Parasyrphus punctulatus ×6

Parasyrphus nigritarsis

Wing length: 8·25–9·25 mm

Identification: Very difficult to recognise as many of the characters are comparative and the shape of the markings on T2 is not as strongly defined as a golf-club as in other *Parasyrphus*. It has moustache-shaped bands on T3 and T4 (see *page 143*), a completely yellow face (usually a dark central band in other *Parasyrphus*), squamae with a bare upperside and legs that are mainly orange with contrasting black tarsi. There is also slight beading along the margins of the tergites.

Similar species: Confusion with *Syrphus* (*pp. 132–135*) is most likely as the abdominal markings are very similar, and there is also slight beading along the margins; although *P. nigritarsis* often has a stronger orange hue than *Syrphus* the colour tends to fade with age. The black tarsi are a strong pointer to this species but should not be considered wholly reliable, as some *Syrphus* (especially *S. torvus* (*p. 134*)) have similarly darkened tarsi. The presence of hairs on the upperside of the squamae is the most reliable feature differentiating *Syrphus* from *Parasyrphus*. The combination of completely orange tibiae and black tarsi should ensure separation from banded *Epistrophe* (*pp. 136–141*), *Eupeodes* (*pp. 146–151*) and also *Megasyrphus erraticus* (*p. 131*) (in which R$_{4+5}$ is strongly dipped).

J F M A M J J A S O N D

Observation tips: A northern and western species that was once regarded as a rarity but is now known to occur far more widely towards south-east England. In Ireland it appears to be highly localised. The larvae feed on leaf-beetle larvae, seemingly most frequently on the larvae of the Green Dock Beetle on dock leaves (see *page 42*), but also on beetle larvae on Alder and willows along rivers and streams. Adults can be found in such places basking on sunlit leaves, but most recent records have been of the highly distinctive larvae and even of eggs laid on Green Dock Beetle egg-masses.

Parasyrphus vittiger

Wing length: 6·25–8·75 mm

Identification: A yellow-banded species that has a dark central marking on a yellow face. Its thorax is slightly dulled by dusting and the legs are mainly dark. The hind tibiae have a dark central ring and pale ends (see *opposite*). Identification only possible from photographs if both the face and legs are depicted.

Similar species: Easily confused with *Syrphus* (*pp. 132–135*), which have a completely yellow face, and some banded *Eupeodes* (*pp. 146–151*), which have short black hairs on the tergite margins. See *opposite* for separation from *Parasyrphus annulatus*, *P. lineola* (N/I) and *P. malinellus*.

Observation tips: Found in conifer plantations and on heathland with conifers. Adults visit yellow flowers, especially yellow Asteraceae. Primarily a northern species, but also widespread on the heathlands of southern England. There is a single post-2000 record from Ireland (in 2017).

J F M A M J J A S O N D

NOTE: This species is listed as *P. relictus* in the European Red List.

144

SYRPHINI: *Parasyrphus*

Parasyrphus hind legs

P. nigritarsis

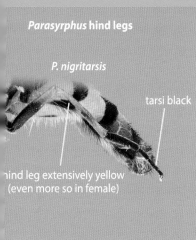

tarsi black

hind leg extensively yellow (even more so in female)

P. vittiger

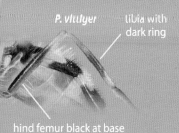

tibia with dark ring

hind femur black at base

P. annulatus

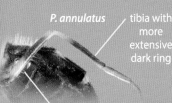

tibia with more extensive dark ring

hind femur yellow at base

P. malinellus

hind leg almost all black except for broadly yellow 'knee'

Note: *P. lineola* (N/I but see *p. 142*) is even blacker and the yellow at the 'knee' is more restricted.

♀ *Parasyrphus nigritarsis* × 6

♂ *Parasyrphus vittiger* × 6

SYRPHINI: *EUPEODES*

> Yellow-and-black banded hoverflies in the genera *Epistrophe, Eupeodes, Syrphus, Parasyrphus* and *Megasyrphus* are consistently misidentified. See pages 128–130 for a guide to their identification.

from 5 p.99

Eupeodes

9 British and Irish species (4 illustrated)

Among the yellow-and-black hoverfly genera, *Eupeodes* can be recognised by the edges of the abdomen, which from abdomen segment T3 onwards have no yellow hairs, only short black ones. The eyes are bare, unlike some other genera such as *Scaeva* (p. 126) and *Dasysyrphus* (p. 120) that look superficially similar. Many of the species are very challenging to identify and there is considerable variation in the colour, shape and extent of markings. This is associated with the temperatures at which the larvae developed. Larvae feed on conifer aphids, but those of the two commonest species, *E. corollae* and *E. luniger*, attack a wide range of ground-layer aphids.

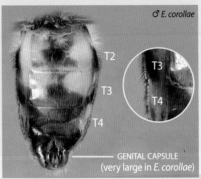

♂ *E. corollae*

GENITAL CAPSULE (very large in *E. corollae*)

Eupeodes abdomen viewed from below showing dark hairs on the edge of segments T3 & T4.

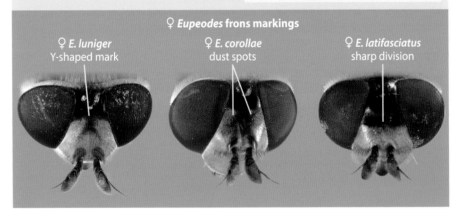

♀ *Eupeodes* frons markings

♀ *E. luniger* — Y-shaped mark   ♀ *E. corollae* — dust spots   ♀ *E. latifasciatus* — sharp division

Eupeodes species not otherwise covered:

Note: *Eupeodes* is a particularly difficult genus, partly because the taxonomy is still in a state of flux, but also because they show a great deal of variation in colour and markings depending on the temperature at which the larvae have developed.	
E. nielseni NS	Two scarce species that are difficult to identify and require microscopic examination of subtle features. *E. nielseni* is mostly recorded from Scotland; *E. nitens* in southern England, north to the midlands and Wales.
E. nitens NS	
E. bucculatus	*E. goeldlini* is a recently recognised split from *E. bucculatus* (probably what is referred to as 'species B' in Stubbs & Falk (2002)). These are very difficult species to identify.
E. goeldlini	
E. lundbecki	On a wider European scale, *Eupeodes* can be difficult to separate from *Scaeva*, and *E. lundbecki* looks more like a *Scaeva* species. It is a vagrant that has been recorded about ten times around the British coast.

SYRPHINI: *Eupeodes*

Eupeodes luniger

Wing length: 6·5–10·0 mm

GB: Widespread =
Ir: Frequent

Identification: Size, markings and coloration are strongly affected by temperature during larval and pupal development (see *page 28*). Most have distinct spots but these can be partially or wholly fused. Both sexes have distinctive bare areas on the wings, especially on the second basal cell, which is largely (but not completely) bare. Females have a distinctive Y-shaped mark on the frons (see *opposite*), formed by dusting, but this character is often poorly depicted in photographs due to reflection from dusting. Identification of males uses a combination of features including the shape and extent of the black area on the abdomen underside S2 and the angle between the eyes (which is very difficult to judge); microscopic examination of wing microtrichia is a far more reliable way of making a firm identification.

Similar species: Males cannot be identified reliably from photographs. Confusion with other *Eupeodes*, especially *E. corollae* (*p. 148*), is frequent because abdominal markings are often difficult to interpret. Some male *E. latifasciatus* (*p. 148*), in which the markings are spots, may also cause problems (check the wings, which are almost entirely covered by microtrichia in that species). Large individuals could be confused with *E. lundbecki* (N/I) and *Scaeva selenitica* (*p. 126*) in which the frons is more prominent and inflated. *E. luniger* males (especially) can be mistaken for *Parasyrphus punctulatus* (*p. 142*), which has longer hairs on the margins of the abdomen and more pronounced golf-club-shaped markings. Confusion also happens with the *Dasysyrphus venustus* complex (*p. 122*) which has a very long and dark stigma.

Observation tips: Partially migratory and widespread, like *E. corollae*, but seems to have a more permanent resident population. It emerges as early as late January. Peak abundance usually occurs in late summer and it may persist through the autumn until the first frosts.

Eupeodes luniger × 6

Eupeodes corollae

Wing length: 5·0–8·25 mm

Identification: An extremely variable, sexually dimorphic species in which the yellow markings are quite broad where they reach the edge of the abdomen segments. These markings occupy 50% or more of the margin in males (25% or less in other species) and this is the only species of *Eupeodes* where they reach the margins in females. Males also have an unusually large and obvious genital capsule (see *page 146*).

Similar species: Most often confused with *Eupeodes luniger* (*p. 147*), in which the abdominal markings do not extend to the tergite margins; females of that species also have a Y-shaped marking the frons (weak dark dust spots on *E. corollae* – see *page 146*). Confusion with other *Eupeodes* is possible but males should be distinctive as they have a very large genital capsule. Females require much greater care, checking both frons markings and the shape of vein R_{4+5}.

J F M A M J J A S O N D

Observation tips: A migratory species whose numbers peak in midsummer. As a consequence, it is likely to be found almost anywhere, although its frequency declines northwards. A very common visitor to garden flowers.

Eupeodes latifasciatus

Wing length: 6·5–8·5 mm

Identification: Females are readily distinguished from other *Eupeodes* as the markings on the abdomen form bands, and there are no dust spots on the distinctive half-black, half-yellow frons (see *page 146*). Males are very variable, their abdominal markings ranging from bands through to lunulate spots. The most reliable character requires a microscope: the 2nd basal cell of the wing is completely covered with microtrichia (75% or less in other species).

Similar species: Males in which the abdominal markings are fully or partially separated may be confused with *Eupeodes luniger* (*p. 147*), requiring examination of wing microtrichia. Although females may be confused with *Syrphus* (*pp. 132–135*) and some other *Eupeodes*, the frons markings are relatively distinctive.

Observation tips: A widespread species which occurs in a variety of situations, perhaps favouring damper woodland edges and meadows. Abundance varies considerably from year to year, suggesting that it is at least a partial migrant.

J F M A M J J A S O N D

Eupeodes larvae lift their prey clear of the plant surface, a behaviour that minimises the dispersal of alarm pheremones released by the aphid to alert its neighbours of danger.

SYRPHINI: *Eupeodes*

Eupeodes corollae × **6**

♂ ♀

GENITAL CAPSULE large

Eupeodes latifasciatus × **6**

♂ ♀

FRONS extended shiny black base; no dusting

Eupeodes lapponicus [LC]

GB: Local [no trend data]
Ir: Rare

Wing length: 8·5–10·0 mm

Identification: A very variable species, with bare eyes and lunulate spots. The variation in size and appearance may arise from differing provenances (locally bred or migrants). The abdominal markings are separated from the margin and do not appear to fuse (as can be the case with other *Eupeodes*). Vein R_{4+5} is strongly dipped, giving the cell beneath (R_{4+5}) a club-shaped appearance.

from 4a p. 99

Similar species: *Scaeva selenitica* (*p. 126*), which has hairy eyes and a more swollen frons. Stubbs & Falk (2002) include 'species A' under the same subgenus, *Lapposyrphus*, but no conclusive proof has emerged that this is a distinct taxon.

Observation tips: Mostly likely to be found in coniferous or mixed woodlands. Seen very infrequently, but increasingly reported by photographers, suggesting that it may previously have been overlooked. In most places it is likely to have been a migrant but there are locations that yield regular records, suggesting some local populations. In Ireland, known from just two old records.

J F M A M J J A S O N D

from 5 p. 99

Meligramma

3 British and Irish species (all illustrated)

Narrow-bodied hoverflies that are similar to *Melangyna* (some authors include them in that genus). They tend to be small and have a yellow scutellum and a yellow face that lacks any dark central 'knob'. They can look rather like *Platycheirus* (pp. 75–93) although that genus has a black face and dark scutellum. *Melangyna* (pp. 154–159) is also similar but the face is only partially yellow and has a darkened central 'knob'. The larvae feed on aphids on trees and shrubs; *M. guttatum* has been reared on aphids on Sycamore.

Meligramma euchromum [LC]

Nationally Scarce
GB: Local ▽
Ir: Not recorded

Wing length: 6·75–8·0 mm

Identification: Somewhat larger and stouter than other *Meligramma* species (*p. 152*). The markings on the abdomen are set at a slight slant that gives a distinctive appearance. This appearance is complemented by a tiny orange spot at the outer corner of the posterior margin of abdomen segments T2 and T3 (although these are often difficult to discern).

Similar species: Confusion may be possible with some *Melangyna* (*p. 154*), in which the abdominal markings are not obliquely angled and the face has at least some black on it – but these species lack the tiny orange spots on abdomen segments T2 and T3 found in *Meligramma euchromum*.

Observation tips: This species is primarily southern and occurs for a comparatively short spell in late April and May. Although it may be found at flowers, it is far more commonly found sun-basking on the leaves of many trees, especially Sycamore, limes and Horse-chestnut.

J F M A M J J A S O N D

NOTE: This species is named *Epistrophella euchroma* in Bot & Van de Meutter (2023).

SYRPHINI: *Eupeodes* | *Meligramma*

♀ *Eupeodes lapponicus* ×6

R_{4+5} strongly dipped

♀ *Meligramma euchromum* ×8

tiny orange spot at the corner of the rear margin of abdomen segments T2 & T3

SYRPHINI: *MELIGRAMMA*

LC *Meligramma trianguliferum*

GB: Frequent ▽
Ir: Rare

Wing length: 5–8 mm

Identification: A narrow-bodied hoverfly with a yellow face and with yellow spots on the abdomen. The markings on abdomen segment T2 can be very small and are near triangular. It is a somewhat cryptic species that might be overlooked among other narrow-bodied hoverflies.

Similar species: Possibly some *Platycheirus* (pp. 75–93), especially more yellowish-marked male *P. albimanus* (p. 84), but all *Platycheirus* have a dark scutellum.

Observation tips: A species of woodland edge, hedgerows and scrub, which is most frequently found in spring, often in late April or May. Although it does visit flowers, such as those of Field Maple, it is more frequently seen basking on sunlit leaves. The larvae of this species is a spectacular and very realistic bird-dropping mimic (see below and *page 17*). This is mainly a species of southern and eastern England, although there are a few scattered records in Scotland and it is possible that its distribution may extend farther northwards in response to climate change. It is known from just two records in Ireland.

J F M A M J J A S O N D

The larvae of *M. trianguliferum* are striking bird dropping mimics that are reported almost as frequently as the adult. They can be found on aphid-infested Elderberry bushes.

LC *Meligramma guttatum*

Nationally Scarce
GB: Local ▽
Ir: Rare

Wing length: 5·25–7·0 mm

Identification: A small, narrow-bodied hoverfly with yellowish or whitish spots on the abdomen and a yellow face without a black facial 'knob'. The markings on abdomen segment T2 can be very small and are indistinctly triangular; those on segments T3 and T4 are rounded, making it distinctly different from other species in the genus. Tends to have yellow spots at the back of the thorax.

Similar species: Superficially some *Platycheirus* (pp. 75–93), although species in that genus have a dark scutellum.

Observation tips: A woodland species which occurs sporadically in low numbers across the country, often along riverbanks and in damp places in southern England, and in Sycamore woods farther north. To date, there have been just seven Irish records. Adults visit a range of flowers including Hogweed and Wild Angelica, mainly in June and July.

J F M A M J J A S O N D

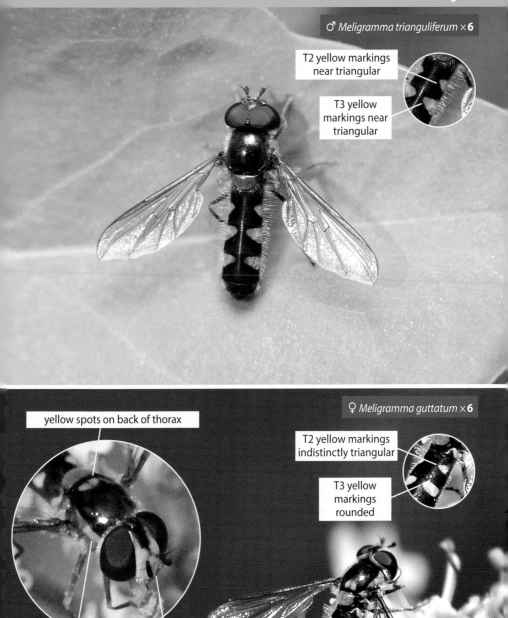

Melangyna

from 5, p. 99

8 British and Irish species (5 illustrated)

This is a challenging genus with several species which are very difficult to identify and others (*M. barbifrons* and *M. ericarum*) that are exceedingly rare. Among the yellow-and-black marked genera with at least partially yellow faces, this genus is relatively small and narrow with mainly black legs. However, some species (especially *M. quadrimaculata*) can have completely black faces, which can be particularly confusing because this genus lies within the Syrphini, most of which have a yellow face. The larvae, where known, feed on a variety of aphids.

Melangyna scutellar markings

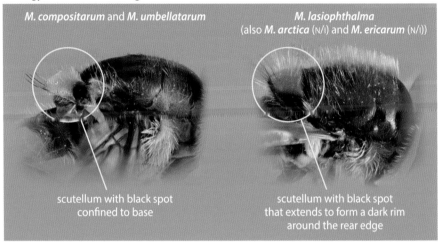

M. compositarum and *M. umbellatarum* — scutellum with black spot confined to base

M. lasiophthalma (also *M. arctica* (N/I) and *M. ericarum* (N/I)) — scutellum with black spot that extends to form a dark rim around the rear edge

Melangyna lasiophthalma

GB: Frequent =
Ir: Frequent

Wing length: 7·0–9·25 mm

Identification: Both sexes have pale hairs on the thorax (longer in males than females) and scutellum. There is a prominent dark marking at the base of the scutellum which extends along the rear edge to form a dark rim (illustrated *above*). In both sexes the abdominal markings reach the tergite margins but in females there is some narrowing that can be misinterpreted (see *Similar species*).

Similar species: *Melangyna compositarum* and *M. umbellatarum* (both *p. 156*) lack a dark rim to the scutellum. *M. arctica/ericarum* (N/I) males have dark thoracic hairs and females have extensive dusting (rather than narrow dust spots). Most likely to be confused with female *Parasyrphus punctulatus* (*p. 142*), but in that species most of the hairs on the scutellum are black (rather than pale), and the eyes of females are bare (rather than weakly hairy).

Observation tips: An early spring species found visiting flowering willows, Blackthorn and other shrubs, but generally flying quite high. Since it is often found with *Parasyrphus punctulatus* and *Melangyna quadrimaculata* (*p. 158*), *Melangyna*-like hoverflies cannot automatically be assumed to be this species! It is widely distributed across Britain and much of Ireland, but possibly disappearing from south-east England; it is seemingly absent from parts of central and north-eastern Ireland.

J F M A M J J A S O N D

SYRPHINI: *Melangyna*

Melangyna lasiophthalma ×**6**

Melangyna compositarum

GB: Widespread =
Ir: Rare

Wing length: 6·25–8·75 mm

Identification: A somewhat narrowly built species with slightly oval markings on abdomen segments T2–4 and a slightly olivaceous thorax that is dulled by fine dusting (see *opposite*). There is a small black spot at the base of the scutellum (see *page 154*). The thoracic and scutellar hairs are pale and the eyes are weakly hairy. In females, the frons is substantially dusted but, again, there is considerable variation. Following Bot & Van de Meutter (2023), the complex of *Melangyna compositarum/labiatarum* is treated here as a single species under the name *M. compositarum*, but it should still be recorded as *M. compositarum/labiatarum*.

Similar species: Confusion with *Melangyna umbellatarum* is most likely as the two species often occur together. In that species the thorax is shinier and relatively free of dusting (but this character is often difficult to interpret in isolation). *M. lasiophthalma* (p. 154), *M. arctica* (N/I) and *M. ericarum* (N/I) have a dark rim to the scutellum in both sexes, whilst males of *M. arctica* and *M. ericarum* have dark thoracic hairs. NOTE: Evaluation of thoracic hair colour requires a view from the side and not from above.

J F M A M J J A S O N D

Observation tips: Occurs in woodland, where it is most abundant from mid-June to mid-August and is a flower visitor, especially to umbellifers (often with *M. umbellatarum*). Males occasionally form small hovering swarms. Known in Ireland from just a single record.

Melangyna umbellatarum

GB: Widespread =
Ir: Local

Wing length: 6·5–8·75 mm

Identification: A somewhat narrowly built species with slightly oval markings on abdomen segments T2–4 and a strongly shining thorax (see *opposite*). The abdominal markings are usually paler than is typical for the genus – often creamy or almost white, but can be distinctly yellow in recently emerged individuals. There is a small black spot at the base of the scutellum (see *page 154*). The thoracic hairs are pale. In females, the dusting on the frons does not reach the antennae, leaving a distinct black undusted band.

Similar species: Confusion with *Melangyna compositarum* is most likely (the two species often occur together) as judging the shininess of the thorax is subjective and dependent upon lighting. Could also be confused with the exceedingly rare *M. arctica* (N/I) but that species has a dark rim to the scutellum and dark thoracic hairs in the male. Recently emerged 'yellow' individuals can be mistaken for *M. lasiophthalma* (p. 154), but in that species the scutellum has a dark rim.

J F M A M J J A S O N D

Observation tips: Found most frequently at Hogweed and other umbellifers in woodland rides or in scrubby grassland, mainly in midsummer, when it can be numerous. Predominantly a southern species in Britain but occurs north into the highlands of Scotland and the Spey Valley. In Ireland it is encountered sporadically in localised areas (mainly) in the south.

SYRPHINI: *Melangyna*

♀ *Melangyna compositarum* × 6

THORAX dull and dusted

Melangyna umbellatarum × 6

THORAX shiny

SYRPHINI: *MELANGYNA*

LC *Melangyna quadrimaculata*

GB: Local ▽
Ir: Rare

Wing length: 7·25–9·0 mm

Identification: A somewhat narrowly built species with a strongly darkened and elongate stigma, a substantially darkened face and frons, and a relatively dark scutellum. Males have spots on abdomen segments T3 and T4; in females the abdomen lacks markings.

Similar species: Most likely to be confused with male *Melangyna barbifrons* (N/I), which has much more weakly hairy eyes and black rather than pale thoracic/scutellar hairs. **Care is needed as these characters are difficult to detect in photographs.** In addition, dark individuals of *Leucozona laternaria* (p. 118) have been misidentified as *M. quadrimaculata* (time of year is a good indicator). The hairs on the face of females may lead to confusion with *Cheilosia variabilis* (group) (p. 178) but the distinctively elongate stigma of *M. quadrimaculata* is a clear separator.

Observation tips. A very early spring species, with a peak flight-period from the end of March to early April. It visits willow flowers in woodland rides (especially in coniferous woodlands), but often flies at a considerable height. Although widely distributed in Britain, the time of emergence means that this species is probably significantly under-recorded. To date, there have been nine Irish records, just one of which is post-2000 (in 2017).

J F M A M J J A S O N D

LC *Melangyna cincta*

GB: Frequent ▽
Ir: Local

Wing length: 6·25–8·75 mm

Identification: One of the more obvious members of the genus, with pointed, triangular markings on abdomen segment T2 and complete bands on the other segments.

Similar species: Most likely to be confused with *Meliscaeva cinctella* (p. 160), which has broad, blunt-ended markings on abdomen segment T2.

Observation tips: A woodland species which, though widespread, is rarely abundant and usually found visiting flowers such as Hogweed. The larvae are known to feed on a variety of tree-dwelling aphids. Although it occurs across much of Britain, it is less common in the north; there are relatively few recent records from Ireland.

J F M A M J J A S O N D

Melangyna species not otherwise covered:

M. arctica	A scarce species of the north and west that seems to prefer conifer woods, although it can sometimes be found in other habitats.
M. barbifrons NT	An extremely scarce, although widespread, woodland species that usually flies early in the spring. Only 15 known records since 1980.
M. ericarum VU	An extremely rare species of the Caledonian pine forest of central Scotland. Only two known records since 1980.

SYRPHINI: *Melangyna*

♀ *Melangyna quadrimaculata* × 6   ♂

♂ *Melangyna cincta* × 6

SYRPHINI: *MELISCAEVA*

from 5
p. 99

Meliscaeva

2 British and Irish species (both illustrated)

Small, elongate, black-and-yellow hoverflies with yellow faces and somewhat elongated wings. At high magnification, the hind margins of the wings can be seen to have a distinct row of small black spots, a feature that they share with *Episyrphus balteatus* (p. 162). They are woodland species which occur in highly variable numbers, especially *M. auricollis* which can, seemingly, barely occur in some years. The larvae feed on aphids on various shrubs and trees.

LC *Meliscaeva cinctella*

GB: Widespread ▽
Ir: Widespread

Wing length: 7·0–9·75 mm

Identification: A relatively long-and-narrow species with a distinctive combination of broad, blunt-ended markings on abdomen segment T2, combined with bands on abdomen segments T3 and T4. The facial knob is yellow and the lunule above the antennae is black.

Similar species: Most likely to be confused with a few other small, banded hoverflies such as *Meliscaeva auricollis* (see table opposite); *Melangyna cincta* (p. 158), which has sharply triangular yellow markings on T2; and possibly *Parasyrphus* (pp. 142–145), especially *P. annulatus* (N/I), which are somewhat broader bodied and have moustachioed bands.

Observation tips: A widespread woodland species which visits a wide range of flowers such as Bramble, ragworts and umbellifers. It often occurs in considerable numbers from June to September.

J F M A M J J A S O N D

LC *Meliscaeva auricollis*

GB: Widespread =
Ir: Widespread

Wing length: 6·0–9·5 mm

Identification: An extremely variable species in which the markings are strongly influenced by the temperature at which larvae develop (see *page 28*). In spring, if development has been in cold conditions, the adults are often darker and less well-marked than in later generations which have developed in warmer conditions. The elliptical markings on abdomen segment T2 are a useful feature. The oblique rear margins of the spots on abdomen segments T3 and T4 are distinctive, but vary hugely and can be fused, forming bands in some. However, the black facial knob and yellow frontal lunule above the antennae are consistent features.

Similar species: *Meliscaeva cinctella*. Individual *M. auricollis* with an almost black scutellum could mistaken for *Platycheirus* spp. within the SCUTATUS group (p. 88) – especially weaker-marked *P. scutatus* (p. 88), which has a black face; *Melangyna cincta* (p. 158), which has sharply triangular, yellow markings on abdomen segment T2.

J F M A M J J A S O N D

Observation tips: Predominantly a woodland edge species. It is one of a small suite of species that can be found throughout the year. Although numbers peak in midsummer, both sexes are frequently observed in midwinter in southern England.

SYRPHINI: *Meliscaeva*

The two *Meliscaeva* species are variable but can usually be separated as follows:

SPECIES	ABDOMEN T2	ABDOMEN T3 & T4	FACE
M. cinctella	MARKINGS broad with blunt ends	MARKINGS bands	FACIAL KNOB yellow; LUNULE ABOVE ANTENNAE black
M. auricollis	MARKINGS usually elliptical	MARKINGS spots with oblique rear margins	FACIAL KNOB typically some black at least LUNULE ABOVE ANTENNAE yellow at front

Meliscaeva faces

M. cinctella — LUNULE ABOVE ANTENNAE black; FACIAL KNOB yellow

M. auricollis — LUNULE ABOVE ANTENNAE yellow at front; FACIAL KNOB black

Beware – the amount of black on the facial knob is variable.

♂ *Meliscaeva cinctella* × 6

T2 yellow markings broad with blunt ends; T3 & T4 banded

dark ♀, spring   *Meliscaeva auricollis* × 6   ♀

T2 yellow markings elliptical; T3 & T4 spots. BEWARE – can be banded, in which case use facial features.

Episyrphus
from 5
p. 99

1 British and Irish species (illustrated)

The commonest hoverfly in Britain and Ireland, commonly known as the Marmalade Hoverfly. This is an extremely variable species whose background colour is highly influenced by the temperature at which the larvae developed (see *page 28*). Larvae in hot conditions produce adults with more orange markings (some can be almost lacking in any black markings), whilst those that develop in cooler conditions produce darker adults (with some completely black). The larvae feed on a wide range of aphid species and can be very numerous in agricultural crops, for example feeding on cereal and cabbage aphids.

LC *Episyrphus balteatus* Marmalade Hoverfly

GB: Widespread ▲
Ir: Widespread

Wing length: 6·0–10·25 mm

Identification: Each abdomen segment has two dark bands separated by two orange bands. This is a feature that is not shared by any other British or Irish hoverfly species.

Similar species: None.

Observation tips: Ubiquitous and very common. Numbers tend to rise through the spring and there is often a marked peak in abundance in late July, but adults occur at all times of year and can even be found on sunny days in midwinter. Mass immigrations from continental Europe occur, sometimes resulting in large numbers of dead hoverflies in the surf on some beaches – and leading to reports in the press of 'plagues of wasps'! Being so abundant means that the larvae of this species are commonly seen in gardens, making them an obvious photographic target.

J F M A M J J A S O N D

eggs — larva — puparium

Watch a female Marmalade Hoverfly *E. balteatus* carefully and you may see her placing her eggs close to an aphid colony. In this example, the eggs have been laid next to aphids on a nasturtium leaf. The larva in the middle photograph is associated with aphids on Himalayan Balsam. The final photograph depicts a puparium; the abdominal markings of the adult hoverfly are clearly visible through the puparial skin, indicating that it will soon emerge. Unlike moth and butterfly larvae, which shed their skin during pupation, those of hoverflies create their pupa within the existing larval skin, hence the term puparium. This arrangement gives the developing hoverfly extra protection against desiccation while it turns from a larva into an adult.

SYRPHINI: *Episyrphus*

E. balteatus is very variable in size and coloration. In spring, adults are often much darker (BOTTOM LEFT) than the summer generation (TOP). Occasionally, very black females occur (BOTTOM RIGHT); it is thought that these may be gynandromorphs (gender uncertain).

♂ summer generation

Episyrphus balteatus × **6**

♀ spring generation

♀ black individual

Guide to Callicerini

from 5a
p.57

Represented by a single genus of charismatic hoverflies, easily recognisable by their **porrect antennae with a white-tipped terminal arista**.

Callicera
3 British species (all illustrated)

Highly charismatic species that are rarely encountered as adults. They are more easily found by searching for larvae, which live in water-filled rot-holes. However, trees with rot-holes are frequently regarded as dangerous – which can lead to them being felled in urban areas or parklands. This is a particular concern in relation to *C. aurata* and *C. spinolae* (the latter being a UK Biodiversity Action Plan (UKBAP) Priority Species (see *page 322*)), which are most likely to occur in such situations. Although many *Callicera* will be readily recognised, it is important to pay attention to the relative lengths of the antennal segments since there are other (non-British) species that might occur. In *C. aurata* and *C. spinolae* antennal segment 3 is about as long as the combined lengths of segments 1 and 2, whereas in *C. rufa* antennal segment 3 is approximately twice the length of the combined lengths of segments 1 and 2 (see *opposite*).

Guide to British *Callicera* species

SPECIES	FEATURES		
C. rufa (below)	FORM distinctive, with foxy fur and base colour **black**		
C. aurata (p. 166)	FORM base colour **bronzy**	LEGS	♀ basal half of femora **black** ♂ **black** hairs between front and middle legs
C. spinolae (p. 166)		LEGS	♀ basal half of femora **orange** ♂ **yellow** hairs between front and middle legs

Callicera rufa

Wing length: 9·75–11·25 mm

Identification: Highly distinctive with a black base body colour, foxy fur and yellow legs – except for the two outermost tarsal segments, which are black. Tends to be smaller than the other two *Callicera* species.

Similar species: Only likely to be confused with other *Callicera* (see table *above*). In case of doubt, the relative length of the antennal segments (see *opposite*) will distinguish *C. rufa* from the others in the genus.

Observation tips: Adults are rarely seen, although males will 'hill-top' and lek around sunlit Scots Pine trunks, often well above head height. The larvae live in rot-holes in Scots Pine and in decaying stumps of a variety of other conifers in felled plantations. Searches for larvae have shown it to be quite widespread in the highlands of Scotland and not as rare as previously thought (see *page 43*). Furthermore, it has been found to colonise artificial rot-holes cut in pine stumps in areas where it was not previously recorded. Together with a dramatic range expansion into England and Wales over the past decade, this evidence suggests that *C. rufa* might occur almost anywhere in England and Wales where there are old Scots Pines. Although most records are for May and June, a recent record of a male in October points to the possibility of a partial late generation in southern England.

J F M A M J J A S O N D

CALLICERINI: *Callicera*

Callicera antennae

C. rufa: antenna segment 3 is about **twice as long** as the combined length of segments 1 and 2.

C. spinolae and **C. aurata**: antenna segment 3 is about the **same length** as the combined length of segments 1 and 2.

♀ *Callicera rufa* ×6

C. rufa larva

Callicera aurata

Wing length: 10·0–12·5 mm

Identification: One of two species (with *C. spinolae*) in which the base body colour is bronzy. Females are best identified by the basal half of the femora, which is black. Males are more difficult to identify but have black hairs between the front and middle legs.

Similar species: Only likely to be confused with other *Callicera* species, especially *C. spinolae*, in which females have entirely orange legs and males have yellow hairs between the front and middle legs (tricky to see).

Observation tips: Most records are from southern England, although there is a recent Scottish record. The larvae live in water-filled cavities in a variety of trees. Adults visit flowers such as Hawthorn, umbellifers and Ivy. Usually flies between June and August, but has occasionally been reported into September, when the flight-period overlaps with that of *C. spinolae*.

J F M A M J J A S O N D

Note: *Callicera spinolae* tends to fly later in the year, but care is needed because *C. aurata* has also been found occasionally in autumn and there is a single record of *C. rufa* (p. 164) flying in October; consequently, the date of observation should not be used as a guide to identification, especially when males are involved.

Callicera spinolae

Wing length: 12–15 mm

Identification: One of two species (with *C. aurata*) in which the base body colour is bronzy. Females are best identified by their orange hind femora and dark bands on the abdomen. Males are more difficult to identify but have yellow hairs between the front and middle legs (tricky to see).

Similar species: Only likely to be confused with other *Callicera* species, especially *C. aurata*, in which females have the basal half of the femora black and males have black hairs between the front and middle legs.

Observation tips: Once known from only a few localities in East Anglia, *C. spinolae* has extended its range quite dramatically into the London suburbs and farther into Kent, Surrey and Sussex. Adults fly in autumn and visit Ivy flowers. As a UKBAP Priority Species, it has been the subject of detailed studies, including extensive searches for larvae. This has shown that it can use remarkably small rot-holes in a variety of tree species. Its current range expansion seems likely to be linked to climate change.

J F M A M J J A S O N D

CALLICERINI: *Callicera*

♂ *Callicera aurata* ×6

♀ *Callicera spinolae* ×6

Females are readily recognised by their orange hind femora and dark bands on the abdomen. The white tips to the antennae are typical of the genus.

from 7a
p.59

Guide to Cheilosiini

There are four genera within this tribe including our largest genus (*Cheilosia*). The humeri are hairy and the head is sufficiently separated from the thorax to make them visible in many cases. **The Cheilosiini can be readily identified by the presence of a zygoma** – a flat margin to the eye, below the antennae, which is clearly separated from the rest of the face by a fold or groove (see *opposite* and *page 170*). Most species are black but a few are more brightly coloured, *e.g. Rhingia (p. 194)*, or are covered in coloured fur, *e.g. Cheilosia illustrata (p. 176)* and *C. chrysocoma (p. 186)*.

1

a) **Colourful species:**
- Abdomen with rectangular grey dust markings (when fresh)
- Face without obvious 'nose' but lower half projecting somewhat
- Antennae orange
  ▶ *Portevinia maculata* p. 192

Beware – some *Platycheirus* [Bacchini] (pp. 75–93) also have grey markings on the abdomen (see *page 84*).

- Thorax with distinct grey stripes; abdomen brassy; wings with dark markings
  ▶ *Ferdinandea* p. 192

- Mouth edge projecting into an obvious 'beak'
- Abdomen orange
  ▶ *Rhingia* p. 194

b) **Less colourful species:**
- Face with distinct 'nose'   ▶

CHEILOSIINI

1a) and 1b) Cheilosiini faces

The presence of the zygoma is unique to the Cheilosiini among British hoverflies.

RIGHT: *Cheilosia vulpina* showing the zygoma and distinct 'nose'.

BELOW RIGHT: *Portevinia maculata* illustrating the zygoma, the lack of an obvious 'nose' and the slight projection of the lower half.

BELOW: *Rhingia rostrata* showing the more extreme 'beaked' appearance of that genus.

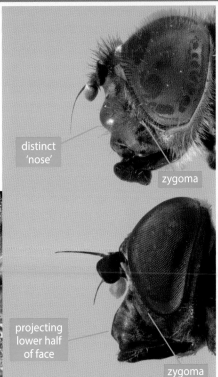

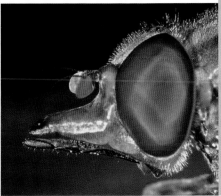

2 **a) Furry species**

from **MINING BEE MIMIC** [good]
1b ▶ *Cheilosia chrysocoma* p. 186

BEE MIMICS [poor]
  ▶ *Cheilosia albipila* p. 184
  ▶ *Cheilosia grossa* p. 184

BUMBLEBEE MIMIC [poor]
  ▶ *Cheilosia illustrata* p. 176

b) Black hoverflies with few or no bristles

Note: This is the region's largest genus and includes species of very varied size and form.
  ▶ *Cheilosia* pp. 170–191

from 2a, 2b
p.169

Cheilosia
38 British and Irish species (17 illustrated)

The genus *Cheilosia* contains the largest number of hoverfly species in Britain and Ireland. These are basically 'black jobs' – although a few are furry. Novices often dismiss them as too difficult to attempt but, in fact, many are actually quite distinctive and have a definite jizz. A feature of the tribe Cheilosiini is the presence of the 'zygoma' – a sharply defined region of the face running along the eye margins (see *below*). As the other genera of Cheilosiini (*Ferdinandea* (*p. 192*), *Portevinia* (*p. 192*) and *Rhingia* (*p. 194*)) have other distinctive features, the zygoma serves as a particularly good character for recognising *Cheilosia*. **The other major group of largely black species is the tribe Pipizini (*p. 254*), which lack a zygoma and do not have a facial prominence or 'nose'; instead, they have a flat, strongly hairy face.**

The larvae of *Cheilosia* feed in the leaves, stems and roots of plants and a few are associated with fungi. Some species have two or more generations per year, but the majority have a single generation that occurs at a very specific time of the year – so the date of capture can be an important identification pointer.

Identifying *Cheilosia*

It takes time to develop the knowledge and experience to recognise *Cheilosia* species in the field and reliable identification of the majority requires careful examination under magnification and good light.

Full identification of all *Cheilosia* species is beyond the scope of this book, since it requires detailed keys and access to voucher specimens for comparison.

Separating males and females
Males and females often have different features, and some species have two broods that differ from one another. In addition, females of some species have characters that are very similar to those of the males of other species – **so it is important first to sex an individual, which is best done based on whether the eyes touch at the top of the head (♂s) or are separated (♀s).**

Identification caveats
- The characters used for identification are often small and subtle, and many are best evaluated in comparison with other species.
- Furthermore, it is often necessary to have assessed a suite of these characters to reach an accurate identification.
- To add to the challenge, in some species these characters can be individually variable. This variability and the comparative nature of characters means that it is not always possible to confirm the identification of a specimen.

Nevertheless, there are rules that can be followed that will help to narrow down an identification, starting with the time of year that the insect is found. If, for example, you think you have found *Cheilosia grossa* in August, you should discount this notion as the species is rarely seen after the end of April and certainly not after the end of May (any records thereafter will be of larvae).

Similarly, check whether the species you think you have found occurs in the area: there are many species that are primarily southern and a few that are mostly or exclusively northern. See the species accounts for such information.

Cheilosia 'morpho-groups'
As the genus is big and complex, its treatment in Britain and Ireland has been to split it into 'morpho-groups' (*i.e.* groups of species that have similar characters that can be used to set them apart from others, represented by a lead species). The choice of 'group leaders' is very subjective and differs between UK and European keys. In an upgrade from the last edition of this WILD*Guide*, an interpretation of the key to the main groups used by Stubbs & Falk, taking into account some of the thinking in van Veen has been included. Unfortunately, this does not work in its entirety because there are some species that are polymorphic and others where spring and summer generations are subtly different. Nevertheless, the system broadly works and is a useful way of making *Cheilosia* understandable. Beware – this is a much bigger genus in Europe and is far more complex! Working through each stage in sequence should help you to get to the approximate species grouping.

CHEILOSIINI: *Cheilosia*

Primary identification features of *Cheilosia*

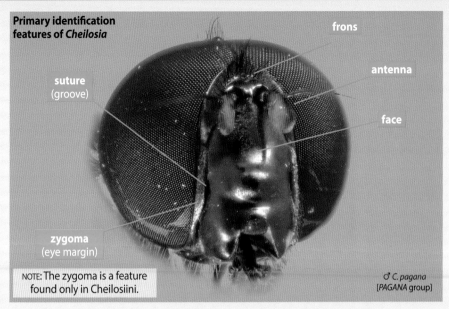

- frons
- antenna
- face
- suture (groove)
- zygoma (eye margin)

NOTE: The zygoma is a feature found only in Cheilosiini.

♂ *C. pagana* [*PAGANA* group]

A step-by-step guide to British and Irish *Cheilosia* species
It is important to work through each stage in sequence.
NOTE: Page numbers in *italic text* refer to the Notes on British and Irish *Cheilosia* table (pp. 174–175).

1 Presence of shading on the wings

NOTE: The following two species have distinctive wing shading but some others, especially *C. albipila* and *C. nebulosa*, have hints of shading, although this shading is far less obvious.

| WINGS with strong disruptive shades; FORM furry bumblebee mimic | WINGS strong infuscation over the main cross-veins; FORM body largely bare but with distinct tufts of pale hairs on the sides of the thorax and abdomen |

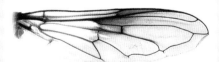 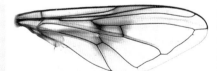

▶ *C. illustrata* p. 176 | p. 174   ▶ *C. caerulescens* p. 176 | p. 174

CHEILOSIINI: *CHEILOSIA*

It is important to work through each stage in sequence.
NOTE: Page numbers in *italic text* refer to the Notes on British and Irish *Cheilosia* table (*pp. 174–175*).

2 Presence/absence of facial hairs

Beware – this feature is often almost impossible to detect in field photographs, making it very tricky to use. Several species in this group could easily be confused with others that lack facial hairs, and there are a few in which the hairs are minute and difficult to detect even under the microscope (*e.g. C. latifrons*).

FACE **with erect, projecting hairs**   FACE **bare**

 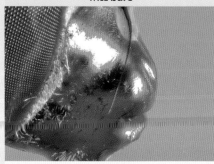

VARIABILIS group [6 spp.; 2 ILL.]
▶ *p. 178* | *p. 174*

▶ 3

Beware – *Melangyna quadrimaculata* (*p. 158*) ♀s may be mistaken for this group [check for zygoma].

3 Presence/absence of eye hairs

Beware – ♂s (eyes touching) of the *ALBITARSIS* group have hairy eyes and ♀s (eyes separated) have bare eyes although well-marked individuals can be told by the leg (tarsal) coloration without the need to check eye hairs.

EYES **bare**   EYES **with hairs**

▶ 4   ▶ 5

***Cheilosia* leg colour**

This can be very variable and requires careful interpretation. There are, however, four basic groups:

(A) all legs entirely black;
(B) all legs entirely black apart from paler tarsal segments 2–4 on the front and middle legs;
(C) femora and tibiae partly black but with orange/yellow areas, mainly at the apices of each section but also on the tarsi;
(D) hind tibiae completely orange although a faint smudge of darkening can be present.

The combination of eye hairs presence/absence and leg colour should enable the assignment of an individual to one of the following morpho-groups.

CHEILOSIINI: *Cheilosia*

4 EYES bare

Beware – some individuals of the *VARIABILIS* group may fall in this category.

(A) LEGS all entirely black

ANTIQUA group
[6 spp.; (N/I)] ▶ *p. 174*

Beware – *ALBITARSIS* group ♀s with darkened tarsi (*p. 190*).

(B) LEGS black with paler tarsal segments 2–4 on front and middle legs

ALBITARSIS group [3 spp.; 2 ILL.]
▶ *p. 190* | *p. 174*

Typical black legs + pale tarsi of the *ALBITARSIS* group are distinctive

(C) LEGS mixed black and yellow/orange

PAGANA group
[4 spp.; 4 ILL.]
▶ *p. 180* | *p. 174*

5 EYES with hairs

(A) LEGS all entirely black

▶ *C. impressa p. 190* | *p. 175*

Beware – *ALBITARSIS* group ♀s with darkened tarsi (*p. 190*).

(B) LEGS black with paler tarsal segments 2–4 on front and middle legs

ALBITARSIS group (♂s)
[3 spp.; 2 ILL.]
▶ *p. 190* | *p. 174*

(D) LEGS hind tibiae completely orange

GROSSA group
[4 spp.; 4 ILL.]
▶ *p. 184* | *p. 175*

(C) LEGS mixed black and yellow/orange

This section comprises three morpho-groups, separation of which depends upon determining whether the abdomen underside segments (sternites) are dusted or shiny. As this character is unlikely to be seen by the general user of this book, the three groups must be treated with caution, especially as some species fall into more than one group because their antennal colour ranges between orange and black!

ABDOMEN UNDERSIDE SEGMENTS dusted		ANTENNAE orange or black	*PROXIMA* group [3 spp.; 1 ILL.] ▶ *p. 188*	*p. 175*
ABDOMEN UNDERSIDE SEGMENTS shiny		ANTENNAE orange or black	*CARBONARIA* group [4 spp.; (N/I)] ▶ *p. 175*
		ANTENNAE orange	*BERGENSTAMMI* group [3 spp.; 1 ILL.] ▶ *p. 188*	*p. 175*

CHEILOSIINI: *CHEILOSIA*

Notes on the British and Irish *Cheilosia* species

WINGS WITH A CLOUD OR DISTINCT INFUSCATION ACROSS THE CROSS-VEINS

Distinctive species with a wing cloud or distinct infuscation across the cross-veins	
C. illustrata p.176	Bumblebee mimic
C. caerulescens p.176	Body largely bare; thorax and abdomen sides with pale hair tufts

WINGS WITHOUT A CLOUD OR DISTINCT INFUSCATION 1/2

FACE with erect projecting hairs	
VARIABILIS group	
C. variabilis p.178	Separated on the basis of antennal colour, presence/absence/colour of scutellar marginal bristles, and face shape. Often very tricky to separate and dependent upon comparison with voucher specimens. Only *C. latifrons* is known from Ireland, although until the taxonomic status of *C. griseiventris* is resolved it is not clear whether that species is also present
C. vulpina p.178	
C. barbata NS	
C. griseiventris	
C. lasiopa	
C. latifrons	

FACE with no erect projecting hairs	1/2
EYES variable	Note: ♀s have bare eyes; ♂s have hairy eyes.
ALBITARSIS group	LEGS black, except for paler SEGS. 2–4 of front and middle tarsi
C. albitarsis p.190	*C. mutabilis* is very much smaller and narrower than *C. albitarsis* agg. (*C. albitarsis* and *C. ranunculi*) and has black halteres (orange-tipped in *C. albitarsis* agg.). **Beware** – *C. mutabilis* can have paler parts to the legs and is confusable with *C. urbana* (and vice versa) if using van Veen (2004). There are scattered records from across Britain but it is absent from Ireland.
C. ranunculi p.190	
C. mutabilis NS	
EYES bare	
ANTIQUA group	LEGS black
C. ahenea VU	♂'s and ♀s are separated on different characters involving relative hair lengths on the thoracic dorsum and the abdomen, together with dusting on the frons and facial profile. All are very difficult to separate without reference to voucher specimens. With the exception of *C. nigripes*, which is southern, they are all primarily or exclusively northern and western in distribution. *C. nigripes* and *C. sahlbergi* do not occur in Ireland and there are only pre-2000 records of *C. ahenea* from Ireland.
C. antiqua	
C. nigripes NS	
C. pubera NS	
C. sahlbergi VU	
C. vicina	
PAGANA group	LEGS black and orange
C. pagana p.180	Separated using a combination of features including antennal colour, facial profile (view from above and side), subtle differences in tarsal colour. There is some potential for confusion between *C. soror* and *C. scutellata* as antennal colours vary between individuals.
C. soror p.180	
C. scutellata p.182	
C. longula p.182	

CHEILOSIINI: *Cheilosia*

WINGS WITHOUT A CLOUD OR DISTINCT INFUSCATION 2/2
FACE with no erect projecting hairs 2/2
EYES hairy

IMPRESSA group

C. impressa p. 190	The entirely black legs are a useful indicator but beware as female *C. albitarsis* often get misinterpreted when the front and middle tarsi are darkened (a very common occurrence).

GROSSA group

C. grossa p. 184	Separated on the presence or absence of scutellar marginal bristles, dusting on the abdomen underside segments (sternites) and hairs on the hind femur (some ♂s). Sexual dimorphism in this group means that ♂s and ♀s can look quite different. *C. nebulosa* resembles a small *C. albipila* and ♀s can occasionally be recognised from photos on the basis of hind tibia colour (*C. nebulosa* mainly black; *C. albipila* orange). *C. nebulosa* is widespread but rarely seen in Britain; *C. uviformis* is known from a very few widely dispersed records (but none post-2000). Both species are known from Ireland but there are few records.
C. albipila p. 184	
C. chrysocoma p. 186	
C. fraterna p. 186	
C. nebulosa	
C. uviformis	

BERGENSTAMMI group

C. bergenstammi p. 188	All have hairy eyes and strongly bicolored tibiae and tarsi. Compared with *C. bergenstammi*, *C. psilophthalma* and *C. urbana* are comparatively tiny species which are separated based on the colour of the tarsal claws. *C. psilophthalma* has been recorded from Ireland (four pre-2000 records) but *C. urbana* has not.
C. psilophthalma `DD`	
C. urbana	

CARBONARIA group

C. carbonaria `NS`	♂s and ♀s are separated using different characters involving the facial profile, abdomen-shape, dusting, and the shape of the grooves (carinae) on the frons (*C. vernalis/cynocephala*). Only *C. vernalis* is widespread; the rest are extremely rarely encountered, although *C. semifasciata* could become more widely distributed if introduced with imported Orpine plants. Only *C. semifasciata* and *C. vernalis* occur in Ireland.
C. cynocephala `NS`	
C. semifasciata `NT`	
C. vernalis	

PROXIMA group

C. proxima p. 188	Separated by facial profile, colour of hairs on the top of the thorax, scutellum and abdomen, and antenna colour. **All are highly subjective characters that require comparison with voucher specimens.** *C. velutina* is found primarily south of a line between the Mersey and the Humber, but with scattered records to Scotland; there is just a single pre-2000 Irish record. *Cheilosia* 'species B' is thought likely to be *C. gigantea* (a widespread species in continental Europe) but this cannot be confirmed.
C. velutina `NS`	
C. 'species B'	

The *PROXIMA* group is separated from the *BERGENSTAMMI* and *CARBONARIA* groups on the basis of dusting on the abdomen underside segments (sternites), which are dusted in the *PROXIMA* group, shiny in the other two. There is considerable overlap between the *BERGENSTAMMI* and *CARBONARIA* groups and a different approach is therefore needed. Some species (e.g. *C. vernalis*) show individual variation that can fall into both groups and others (such as *C. urbana*) according to van Veen (2004) can be confused with *C. mutabilis* of the *ALBITARSIS* group. Consequently, species such as this require a highly detailed, specialised approach to their descriptions and treatment in keys – something that is far beyond the scope of this book.

Distinctive *Cheilosia* with wing clouds 2 SPP. | 2 ILL.

Cheilosia caerulescens

Wing length: 6–7 mm

Identification: A stout black *Cheilosia* with clouding across the inner cross-veins (although this is individually variable in extent and density), a distinctively projecting face without erect hairs (although heavily dusted), bare eyes and partially yellow legs. It is quite hairy on the thorax and abdomen. Adults tend to rest with folded wings, which can obscure the wing clouds.

Similar species: *Cheilosia illustrata*, but that species is hairy and a bumblebee mimic. Unless the wing cloud is noticeable, confusion with *C. proxima* (p. 188) and *C. vulpina* (p. 178) may be possible in photographs. The shape and dusting of the face, providing this is visible, should be conclusive in such cases. *C. pagana* (p. 180) also looks similar but has orange antennae.

Observation tips: The larvae mine the leaves of houseleeks, *Sempervivum* spp., causing obvious damage and making it easier to record larval mines than the adults. It is double-brooded, with adults flying in the spring and in midsummer. Adults visit a variety of garden flowers. First recorded in Britain in 2006, and likely to have been introduced in imported plants from Europe. Now widespread in urban environments, it occasionally turns up away from gardens. Mainly an English species but increasingly extending into south Wales and with a single record from the Edinburgh area. It may eventually turn up in Ireland.

J F M A M J J A S O N D

Identifying wasp and bee mimics pp. 62–65

Cheilosia illustrata

Wing length: 8·5–10·25 mm

Identification: Although this species is described as a bumblebee mimic, it is not a very good one! It has a band of long white hair across the base of the abdomen, dark wing markings, a black scutellum and an all-black face.

Similar species: Most likely to be mistaken for *Leucozona lucorum* (p. 116), but that species has a yellow scutellum. Since it is, at least vaguely, a bumblebee mimic, it could possibly be confused with *Eriozona syrphoides* (p. 116) if the yellow face of that species is not obvious, *Eristalis intricaria* (p. 228), which has a yellow scutellum, and *Volucella pellucens* (p. 272), which has distinctive wing venation (outer cross-vein re-entrant). The only other *Cheilosia* to show a pronounced wing cloud is *C. caerulescens* (but that species is not furry and has a distinctively projecting face).

Observation tips: Strongly associated with Hogweed: larvae mine the stems and roots, and adults are frequently seen visiting the flowers. However, other umbellifers are also visited and adults can therefore be encountered outside the flowering period of Hogweed. It is common and widespread in lowland localities such as woodland edges and road verges, but also found in the uplands where Hogweed occurs.

J F M A M J J A S O N D

CHEILOSIINI: *Cheilosia*

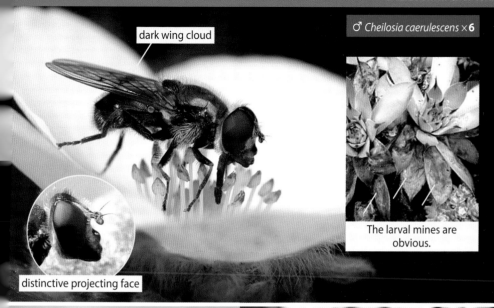

♂ *Cheilosia caerulescens* × 6

dark wing cloud

distinctive projecting face

The larval mines are obvious.

Cheilosia illustrata × 5

dark wing cloud

Cheilosia with erect, projecting hairs on the face

VARIABILIS group   6 SPP. | 2 ILL.

Cheilosia variabilis

Wing length: 7·75–10·25 mm

Identification: A large black *Cheilosia* with a long body, even longer wings and completely black legs. Adults often rest with their wings slightly parted, giving it a delta-winged look. To be sure of the identification, it is important to check for distinct erect, projecting hairs on the sides of the face below the 'nose'.

Similar species: *Cheilosia albitarsis/ranunculi* (p. 190) males, in which some of the tarsal segments are orange. All other species with a hairy face have partially yellow legs, and in some cases the underside of the 3rd antennal segment is reddish. Female *Melangyna quadrimaculata* (p. 158) is also black with long wings but has a much longer and more obvious stigma and longer hairs on the scutellum.

Observation tips: Typically found in damp woodland, streamsides and shady road verges. Both sexes like to bask on sunny leaves, often on low-growing vegetation. The larvae feed in the stems and roots of Common Figwort. This species seemingly occurs in Britain wherever Common Figwort is present, although it is scarcer in the north and absent from parts of Scotland. In Ireland, it appears to be mainly a southern species but should be looked for elsewhere, as its host plant is much more widely distributed.

J F M A M J J A S O N D

Cheilosia vulpina

Wing length: 7–10 mm

Identification: A combination of distinct erect, projecting hairs below the 'nose', together with distinct yellow markings on the dark legs point towards this species. Females exhibit distinct hair stripes on the abdomen that are similar to those of female *C. proxima*.

Similar species: Confusion with *Cheilosia* species with dark antennae, such as *C. variabilis*, *C. proxima* (p. 188) and *C. velutina* (N/I), is most likely, but also with other species that have erect, projecting hairs on the face such as *C. lasiopa* (N/I) and *C. griseiventris* (N/I). Can be confused with female *Melangyna quadrimaculata* (p. 158) which similarly is black with long wings and has a hairy face – the time of year should be enough to separate the two in southern England but if in doubt check the length of the scutellar hairs (much longer in *M. quadrimaculata*).

Observation tips: Once regarded as a relatively scarce species, it has now become one of the commoner *Cheilosia* that visit umbellifers in wooded locations, especially in southern England. It is spreading northwards, especially in north-east England, and has been recorded north of the central lowlands of Scotland. There are two generations in southern England: small numbers occur between April and June followed by much bigger numbers in July and August. Flight times in the north are as yet unclear.

J F M A M J J A S O N D

CHEILOSIINI: *Cheilosia*

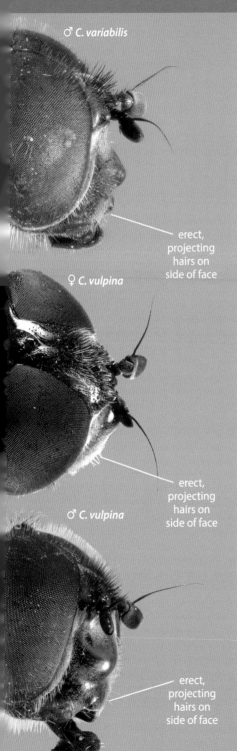

♂ *C. variabilis*

♀ *C. vulpina*

♂ *C. vulpina*

erect, projecting hairs on side of face

erect, projecting hairs on side of face

erect, projecting hairs on side of face

♂ *Cheilosia variabilis* × **6**

tends to rest with wings held in delta shape

♀ *Cheilosia vulpina* × **6**

yellow markings on legs

Cheilosia pagana

Wing length: 4.75–8.5 mm

Identification: This is a distinctly shiny black hoverfly of variable size, with orange antennae and hairless eyes. Males are slightly more difficult to separate from other *Cheilosia* than are females, although the orange antennae with black tips provide a good indication. The antennae of females tend to be disproportionately large, making them rather more distinctive.

Similar species: Confusion is most likely with *Cheilosia soror* and other *Cheilosia* with orange antennae – *C. bergenstammi* (p. 188) and *C. fraterna* (p. 186) – although both of those species have hairy eyes. *C. caerulescens* (p. 176) looks similar but has dark wing clouds and dark antennae.

Observation tips: A common and widespread multi-brooded species which occurs throughout the season. It can be found at Lesser Celandine early in the spring, at Cow Parsley in early summer and at Ivy in the autumn. Larvae have been found among the rotting roots of Cow Parsley. Whilst widespread across much of Britain and Ireland, it is scarcer in the north.

GB: Widespread =
Ir: Frequent

J F M A M J J A S O N D

Cheilosia soror

Wing length: 6–9 mm

Identification: Considerable size range. The eyes are bare, the arista is pubescent, and there are distinctive pits on the antennae of the female (visible on the inner face under high magnification). In profile, and from above, the facial prominence is clearly bulbous. In the female, the scutellum is often (but not always) yellow-tipped; the combination of orange antennae and yellow-tipped scutellum is diagnostic for females.

Similar species: Other bare-eyed *Cheilosia* (see *page 173*). Males can be tricky to separate from *C. pagana* (which usually have a slight black tip to the antennae) and from some *C. scutellata* (p. 182) when the antennae are darkened. Females can also be mistaken for *C. pagana* but have smaller antennae with distinctive pits. The yellow tip to the scutellum is a feature shared with *C. longula* (p. 182) and *C. scutellata* but both of those species have dark antennae.

GB: Frequent ▲
Ir: Not recorded

J F M A M J J A S O N D

Observation tips: A frequent visitor to Hogweed, Wild Angelica and other umbellifers. Once thought to be associated with truffles, its recent dramatic range expansion suggests that a wider variety of fungi is utilised. Formerly confined to southern regions with calcareous influences, south of a line between The Wash and the Severn Estuary, it is spreading northwards, and is now the commonest *Cheilosia* species in many places in southern England in midsummer.

NOTE: This species is named *Cheilosia ruffipes* in Bot & Van de Meutter (2023).

CHEILOSIINI: *Cheilosia*

♀ *Cheilosia pagana* ×6

♂ ANTENNAE orange with black tip

♀ ANTENNAE distinctively large

♀ *C. soror*

ARISTA with strong pilosity

ANTENNA SEGMENT 3 with pronounced pits

somewhat bulbous facial prominence

Cheilosia soror ×6

Cheilosia **with no projecting hairs on the face | EYES bare | LEGS partially pale**   **2/2**
PAGANA group 2/2   **4 SPP. | 4 ILL.**

LC *Cheilosia scutellata*

GB: Frequent =
Ir: Scarce

Wing length: 6–9 mm

Identification: A fairly large and somewhat elongated *Cheilosia* with bare eyes, dark antennae and a very distinct bulging 'nose'. The front tarsi are pale. Females are often distinctive because the tip of the scutellum can be yellow and there can be yellow markings on the face.

Similar species: *C. longula*, which has much darker front tarsi and a distinctly triangular 'nose' (see *opposite*). In general shape and jizz, the most likely confusion is with *C. soror* (*p. 180*) and possibly large *C. pagana* (*p. 180*), both of which have orange antennae. Confusion is also possible with *C. barbata* (N/I) and perhaps even *C. bergenstammi* (*p. 188*) if the antennal colour is not visible in photographs. In the female, the yellow tip to the scutellum is a feature shared with *C. longula* and *C. soror*, so it cannot be taken as diagnostic without checking the antennal colour and the shape of the facial prominence.

J F M A M J J A S O N D

Observation tips: Mainly a woodland species that visits umbellifers in midsummer. Both sexes are frequent visitors to Hogweed, Wild Angelica and other white umbellifers. The larvae feed in *Boletus* fungi. Large numbers of larvae (up to 50) can be found in a single *Boletus* cap and their feeding causes the fungus to liquefy, resulting in a patch of brown slime being all that is left. Whilst widely distributed and sometimes abundant in Britain, this species seems to be confined to small parts of Ireland, a curious distribution that may partly reflect recorder effort.

LC *Cheilosia longula*

GB: Local ▽
Ir: Rare

Wing length: 6·0–8·5 mm

Identification: A comparatively small *Cheilosia* with dark antennae. Viewed from above, the facial prominence is pointed, but this character is often hard to discern in photographs. The front tarsi are predominantly dark. Females often have a yellow-tipped scutellum and yellow markings on the face.

Similar species: *Cheilosia scutellata*, which has pale front tarsi (dark in *C. longula*) and a more bulging 'nose' (see *opposite*). In the female, the yellow tip to the scutellum is a feature shared with *C. scutellata* and *C. soror* (*p. 180*), so it is important to check the antennal colour and the shape of the facial prominence.

Observation tips: A comparatively scarce species that is found mostly in acidic environments, especially in northern and western Britain. It also occurs on some southern heathlands. The larvae are fungal feeders that specialise in *Boletus* fruiting bodies. It is a flower visitor that often occurs on umbels and low-growing flowers such as Tormentil. There are just 11 records from Ireland, suggesting that this species is confined to a very few localities.

J F M A M J J A S O N D

CHEILOSIINI: *Cheilosia*

♀ *Cheilosia scutellata* × **6**

C. scutellata
FACE profile
bulbous

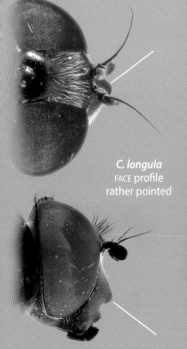

C. longula
FACE profile
rather pointed

♂ *Cheilosia longula* × **6**

Cheilosia with no projecting hairs on the face | EYES hairy 1/4

GROSSA group [HIND TIBIA entirely orange] 1/2   6 SPP. | 4 ILL.

LC *Cheilosia albipila*

GB: Frequent ∨
Ir: Local

Wing length: 8·75–10·75 mm

Identification: Females are distinctive due to the combination of entirely orange hind tibiae, orange antennae and orange fur. The wings have some shading. Males are more problematic and require microscopic examination (see *Similar species*).

Similar species: *Cheilosia nebulosa* (N/I) is smaller with slightly more pronounced wing shading; females have largely black hind femora, and males have short hairs on the hind femora (some considerably longer ones in *C. albipila*). *C. grossa* is similarly sized, normally with orange hind tibiae but black antennae, whereas *C. fraterna* (p. 186) has both the hind tibiae and antennae orange and is less furry.

Observation tips: Flying early in the spring means that this species is often overlooked. Males visit willows and other flowering shrubs, whilst females can be found egg-laying around Marsh Thistle rosettes. The larvae feed in the stems of thistles and, like *C. grossa*, it is more easily recorded, especially in upland areas, by searching for larvae in thistle stems during July (see *page 21*). This species is most frequent in the wetter north and west of Britain, especially on higher ground where Marsh Thistles abound. In Ireland, it appears to be much more localised.

J F M A M J J A S O N D

LC *Cheilosia grossa*

GB: Frequent =
Ir: Scarce

Wing length: 8·5–11·75 mm

Identification: A large, furry hoverfly with orange tibiae and black antennae. This combination of characters makes it hard to confuse with any other *Cheilosia*.

Similar species: Unlikely to be confused with any other species, although individuals with a strong black ring on the hind tibia do occur; such specimens may key out to the PROXIMA group in Stubbs & Falk (2002) or to the CANICULARIS group in van Veen (2004) but they have bristles on the scutellum (absent in the GROSSA group). Some very colourful Scottish forms of *C. grossa* might be mistaken for *C. chrysocoma* (p. 186) but the antennae of that species are orange. Late summer records often prove to be *Eristalis tenax* (p. 224), a misidentification that is difficult to explain.

Observation tips: Can be found visiting willows and other flowering shrubs but is often missed because it flies very early in the year. Males are noted for hovering at or above head height in open ground. It is more readily found during July and early August by searching for larvae feeding in the roots and stems of Marsh and Spear Thistles (see *page 21*). Larval records show that it is much commoner and more widespread than records of adults suggest. Although widespread across Britain, *C. grossa* seems to be genuinely absent from parts of the south-west. It is scarce in Ireland, occurring in just a few areas.

J F M A M J J A S O N D

CHEILOSIINI: *Cheilosia*

♀ *Cheilosia albipila* ×6

Female *C. albipila* egg-laying on Marsh Thistle rosette (see *page 21*).

ANTENNAE orange

HIND TIBIAE orange in both *C. albipila* and *C. grossa*

Cheilosia grossa ×6

ANTENNAE black

| Cheilosia with no projecting hairs on the face | EYES hairy | 2/4 |

GROSSA group [HIND TIBIA entirely orange] 2/2   6 SPP. | 4 ILL.

LC *Cheilosia fraterna*

GB: Frequent ▽
Ir: Not recorded

Wing length: 6·25–10·25 mm

Identification: Males and females show strong sexual dimorphism although both sexes have orange antennae and a wholly orange hind tibia. Males have some short hairs on the underside of the hind femora and a distinct area of black hairs towards the middle and rear of the thorax; females have very short and closely addressed golden hairs in this area.

Similar species: The most likely confusion species is *C. bergenstammi* (*p. 188*), which normally has a dark ring on the tibia but also has much longer scutellar hairs in the male (plus more extensive long hairs on the underside of the hind femora) and much longer thoracic hairs in the female.

Observation tips: Most records are from wet grasslands and upland situations where Marsh Thistles abound. Thistles infested with *C. fraterna* larvae are very distinctive, as they are heavily branched – although beware that Frosted Orange moth larvae also create such deformity. *C. fraterna* is more frequent in the north and west of Britain, where it can be quite common.

J F M A M J J A S O N D

Identifying wasp and bee mimics pp. 62–65

LC *Cheilosia chrysocoma*

Nationally Scarce
GB: Local =
Ir: Local

Wing length: 8·0–10·25 mm

Identification: A highly distinctive mimic of the female of the Tawny Mining Bee (see *opposite*). The body hairs are long and red-orange and the underlying body colour is similar, making it difficult to confuse with any other hoverfly. The antennae are orange. Consequently, this is one of the few *Cheilosia* that can be readily identified from a photograph.

Similar species: Some colour forms of *Merodon equestris* (*pp. 29 and 248*) which can be distinguished by the loop in wing vein R_{4+5} and the presence of a triangular projection on the enlarged hind femur. Some colourful forms of *C. grossa* (*p. 184*) (especially in Scotland) may also cause confusion but that species has black antennae.

Observation tips: A rare species with few, scattered records, often from woodland edges. Early in the year, and in such locations, adults basking on sunny leaves look remarkably like Tawny Mining Bee. Has been seen ovipositing on Wild Angelica, which is believed to be the larval food plant.

J F M A M J J A S O N D

CHEILOSIINI: *CHEILOSIA*   Guide to *Cheilosia*: pp. 170–175

Cheilosia with no projecting hairs on the face | EYES hairy	3/4
BERGENSTAMMI, PROXIMA (and CARBONARIA) groups	10 SPP. | 2 ILL.
[HIND TIBIA orange with a dark ring] 1/1	

LC *Cheilosia bergenstammi*

LC

Wing length: 7·25–9·25 mm

Identification: A relatively large *Cheilosia* with orange antennae and densely hairy eyes, although the degree of hairiness is individually variable. The tarsi are extensively pale and the tibiae are usually orange with a dark ring, although some individuals can have paler legs.

Similar species: *Cheilosia fraterna* and *C. bergenstammi* can show variations in leg colour, making this an unreliable identification feature: other characteristics must therefore be evaluated (see *C. fraterna*, p. 186). *C. albipila* (p. 184), *C. nebulosa* (N/I) and *C. grossa* (p. 184) may be confused at first glance, although the furriness of these species should be distinctive. *C. pagana* (p. 180), *C. scutellata* (p. 182) and *C. soror* (p. 180) may also cause confusion due to their orange antennae, although the eyes of these species are not hairy (but beware that *C. bergenstammi* can have almost hairless eyes). Confusion with *C. albitarsis* agg. (p. 190), *C. carbonaria* (N/I), *C. antiqua* (N/I) and *C. vicina* (N/I) is also possible from photographs that do not show the antennae and leg characters well.

J F M A M J J A S O N D

Observation tips: There are two generations: one in the spring and one in high summer. The spring generation is often found at Dandelion flowers and is generally covered in pollen, whilst the summer generation is commonly found visiting ragworts, the stems and roots of which are mined by the larvae. This is a very widespread species but seems to be declining in parts of south-east England. In Ireland, it is widely distributed but with relatively few recent records.

LC *Cheilosia proxima*

LC

Wing length: 6·25–8·5 mm

Identification: One of a suite of several *Cheilosia* species that look very similar. Like many double-brooded *Cheilosia* there are subtle differences between the spring and summer generations. The combination of hairy eyes, bare face, partially pale legs and dusted underside to the abdomen points towards this species.

Similar species: *Cheilosia velutina* (N/I) and *C. vulpina* (p. 178) are the most likely confusion species. Since *C. velutina* is almost identical to *C. proxima*, apart from a subtle difference in facial profile and having paler antennae, that species can never be ruled out without careful examination. *C. vulpina* is readily separated if the facial hairs can be seen but this can be impossible to tell from photographs.

Observation tips: This is a regular visitor to umbellifer flowers and occurs widely in rough grassland with thistles and along woodland rides; the larvae are stem miners in Creeping Thistle.

J F M A M J J A S O N D

CHEILOSIINI: *Cheilosia*

A ♂ *C. bergenstammi* showing sparsely hairy eyes.

typically densely hairy eyes (many hairs with pollen attached)

♂ *Cheilosia bergenstammi* × 6

tibiae with dark ring

tarsi extensively pale

Cheilosia proxima × 6

♀　　　　　♂

Dusted underside of the abdomen is indicative of *C. proxima*.

Cheilosia with no projecting hairs on the face | EYES hairy	4/4
IMPRESSA group [LEGS entirely black]	1 SP. | 1 ILL.

LC *Cheilosia impressa*

GB: Widespread ▲
Ir: Rare

Wing length: 5.75–8.0 mm

Identification: Females are reasonably distinctive due to their being one of the few *Cheilosia* species that have obvious yellow wing bases. Males are very dark blue-black with obviously red eyes.

Similar species: Female *Cheilosia albitarsis/ranunculi* (*p. 190*) and *Chrysogaster* species (*p. 206*) also have yellow wing bases and the former often have almost entirely dark front tarsi. Dusting on the frons should resolve confusion with *C. cemiteriorum* (*p. 206*) (largely undusted). In *Myolepta dubia* (*p. 217*) the yellow abdomen can often be seen through the wings, although it is a rather longer animal. Lance flies (Lonchaeidae) can be misidentified as this species but they have obvious bristles on the thorax and lack a vena spuria.

Observation tips: A frequent visitor at white and yellow umbellifer flowers along woodland edges and rides. Although most abundant south of the Lake District, it occurs at a few largely coastal locations northwards to the central lowlands of Scotland. In Ireland there are historical records, although it does not seem to have been recorded since 1988.

J F M A M J J A S O N D

Cheilosia with no projecting hairs on the face | EYES ♂ hairy; ♀ bare	
ALBITARSIS group [LEGS black with front + middle tarsi segments 2–4 paler]	3 SPP. | 2 ILL.

LC *Cheilosia albitarsis / ranunculi*

GB: Widespread/Frequent =
Ir: Frequent (*C. albitarsis*)

Wing length: 7.0–9.5 mm

Identification: A pair of very similar species (see *opposite* for differences) often referred to as *C. albitarsis* agg. The species pair is distinctive – with dark antennae and entirely black legs, except for tarsal segments 2–4 of the front and middle legs, which are paler (ranging from yellow to murky orange-brown). The halteres are orange. The eyes of males are hairy; those of females are bare.

NOTE: *C. ranunculi* not recorded in Ireland

Similar species: The Nationally Scarce *Cheilosia mutabilis* (N/I) is very similar, differing in its dark halteres; *C. bergenstammi* (*p. 188*) and *C. fraterna* (*p. 186*), which both have orange antennae and generally paler legs; *C. semifasciata* (picture *p. 22*); *C. carbonaria* (N/I) and *C. vicina* (N/I); and females of bare-eyed species such as *C. antiqua* (N/I) and *C. nigripes* (N/I). Females of *C. albitarsis* agg. often have very indistinct orange tarsal segments and are not infrequently misidentified as *C. impressa*.

J F M A M J J A S O N D

Observation tips: Strongly associated with buttercups; the larvae feed around the roots and adults visit the flowers. *C. albitarsis* occurs throughout Britain and Ireland, whilst *C. ranunculi* is scarce in northern England and there are very few Scottish records (the map shows their combined distributions). Both species occur in damp fields with buttercups and they often occur together at the same site and time. The ecological separation between them has yet to be resolved. ***C. albitarsis* is the only representative of this species pair in Ireland.**

CHEILOSIINI: *Cheilosia*

♀ *Cheilosia impressa* × 6

females have distinctive yellow wing bases

LEGS entirely black

♂ males have red eyes, reminiscent of *Chrysogaster solstitialis* (see *page 206*)

♀ *Cheilosia albitarsis / ranunculi* × 6

♂ *Cheilosia albitarsis* × 6

Front and middle legs black with paler tarsal segments 2–4 that contrast with black segment 5.

Note: The most reliable way of separating *C. albitarsis* and *C. ranunculi* is by microscopic examination of the male genitalia; females cannot be separated.

C. ranunculi
♂ front tarsus segment 5 tapers towards the claws

C. albitarsis
♂ front tarsus segment 5 parallel-sided

Ferdinandea

from 1a p. 168

2 British and Irish species (1 illustrated)

An easily recognised genus in which the thorax has a pair of broad, grey, longitudinal stripes and the abdomen has a somewhat metallic sheen. There are strong black bristles on the sides of the thorax, which is an unusual feature among hoverflies. The larvae live in sap runs.

LC *Ferdinandea cuprea* LC

GB: Widespread ▲
Ir: Frequent

Wing length: 7·5–11·25 mm

Identification: An instantly recognisable hoverfly with a metallic, brassy abdomen and grey stripes running along the thorax. The wings have dark markings, the legs are yellow with strong spines on the tibia and the aristae are black. Unusually for a hoverfly, it has a series of strong bristles on the sides of the thorax.

Similar species: Confusion is most likely with the Nationally Scarce *Ferdinandea ruficornis* (N/I), which tends to be smaller and darker with a blue-black abdomen, lacks strong spines on the tibiae and has red aristae (care needed).

Observation tips: A woodland species that basks on sunlit leaves or on tree trunks. It is also a visitor to flowers, especially yellow composites. Widespread across Britain and Ireland but mainly in the south of both regions; it is rarely seen in abundance.

Ferdinandea species not otherwise covered: J F M A M J J A S O N D

F. ruficornis NS	Few, rather scattered records, mostly from the southern half of Britain, many of which are from Malaise traps; it is possible that the adults stay high up in the canopy. Not known from Ireland.

Portevinia

from 1a p. 168

1 British and Irish species (illustrated)

A dark species with silver-grey abdominal markings. The larvae mine the bulbs and stem bases of Ramsons and adult emergence coincides with the flowering of the plant. Males bask on the sunlit foliage and tend to perch with their wings held in a very characteristic delta shape.

LC *Portevinia maculata* LC

GB: Frequent =
Ir: Local

Wing length: 6·0–8·25 mm

Identification: The orange antennae, blunt facial profile and square, grey abdominal markings make this species relatively easy to identify when found in close association with Ramsons.

Similar species: *Cheilosia* with orange antennae (*e.g.* pp. 180, 188), but those can be separated on the shape of the face. *C. fasciata* is a very similar European Ramsons-associated species, but it has not been found in Britain. *Platycheirus albimanus* (p. 84) has also been confused with this species but it lacks a zygoma.

Observation tips: Confined to locations where Ramsons grows (*e.g.* woodlands, mature hedgerows and large gardens). Where Ramsons is abundant, this hoverfly is usually also abundant. Most records are of males, as females tend to stay low down among the plants and are not so readily seen. An old-woodland species, widespread in Britain but as scarce in Ireland as its favoured habitat.

J F M A M J J A S O N D

CHEILOSIINI: *Ferdinandea* | *Portevinia*

♀ *Ferdinandea cuprea* ×6

THORAX with strong black bristles

♂ *Portevinia maculata* ×6

The shape of the face, orange antennae, and rather square grey markings on the abdomen make this a distinctive hoverfly.

Rhingia

from 6b p.58 and 1a p.168

2 British and Irish species (both illustrated)

An unmistakable genus of dumpy orange hoverflies with a very obvious, exceptionally long rostrum that encloses the proboscis and allows the fly to feed on nectar and pollen in deep flowers, such as Red Campion, which other hoverflies cannot reach. The larvae of *Rhingia campestris* live in fresh cattle dung, but have also been found in other highly enriched wet media such as silage. Its abundance in areas with few cattle, such as East Anglia, suggests that breeding habitats other than cow dung must be significant.

Similar species: Both *Rhingia* species occur together in some places and could be confused with one another (see table *opposite*). Both could potentially be mistaken for red-orange muscid flies (Muscidae | *Phaonia*) or the blow-fly *Stomorhina lunata* (Rhiniidae).

LC *Rhingia campestris*

GB: Widespread =
Ir: Widespread

Wing length: 6·0–9·5 mm

Identification: The commoner of the two *Rhingia* species. Generally dull brown in colour with a dull brown face, scutellum and legs, and with darkened edges to the abdomen.

Observation tips: A widespread species that favours woodland rides and field edges but can be found almost anywhere. It occurs most frequently where there are cattle (as its larvae develop predominantly in cattle dung). A wide range of flowers is visited, including those with deep tubes, such as Red Campion and Bluebell, which other hoverflies cannot utilise. There are normally two generations each year. Being so widespread, peak emergence will vary substantially between southern Britain and the north of Scotland. While it may be seen from mid-April to mid-May in south-east England, in central Scotland it flies in late May and June. The summer generation can be more unpredictable: in south-east England it can fail almost completely during droughts and heatwaves but is often common in Scotland in late July and August.

J F M A M J J A S O N D

LC *Rhingia rostrata*

GB: Frequent ▲
Ir: Not recorded

Wing length: 7·5–9·5 mm

Identification: The less common of the two *Rhingia* species and can be overlooked among the much commoner *R. campestris*. A generally bright-looking hoverfly with a bluish thorax and orange-yellow scutellum, legs and abdomen, which does not have darkened edges.

Observation tips: Once confined to the Weald, the Welsh borders and south Wales, this species has undergone a major range expansion, such that it occurs widely across southern Britain to north Wales and almost as far north as the Scottish borders. It seems to be more strongly confined to woodland than *Rhingia campestris*, but since little is known about its larval biology the reasons for its habitat preferences and sudden spread are not understood (it has been bred from larvae obtained from highly enriched artificial water bodies).

J F M A M J J A S O N D

CHEILOSIINI: *Rhingia*

Rhingia species compared:

SPECIES	OVERALL COLORATION	ROSTRUM	SCUTELLUM & LEGS	ABDOMEN EDGE
R. campestris	generally darker and duller	longer	dull brown	darkened
R. rostrata	generally paler and brighter	shorter	orange-yellow	not darkened

♂ *Rhingia campestris* × 8

dark margins to abdomen segments

HIND TIBIA
partially dark

In both photographs note the long proboscis extending into the flower.

♀ *Rhingia rostrata* × 8

pale margins to abdomen segments

HIND TIBIA
completely pale

from
7b, 7c # Guide to Chrysogastrini
p.59

This is a somewhat heterogeneous grouping of ten genera of small to medium-sized, dark hoverflies. Several genera are distinctive but difficult to find. These include *Brachyopa*, *Myolepta*, *Neoascia* and *Sphegina*. The hairy humeri are not covered by the head and are usually easy to see as the thorax hairs are relatively short by comparison. In several genera, however, these hairs may be small and sparse, and can be bristle-like; as such they are easily overlooked. **The most useful feature for distinguishing this tribe is the shape of the face; unlike many other dark hoverflies it is often strongly concave and does not have a nose-like 'knob'.** However, there are exceptions, especially within *Chrysogaster* and *Melanogaster*. Several genera have quite a strong metallic sheen, especially *Lejogaster*, *Orthonevra* and *Riponnensia*.

1
a) **Abdomen distinctly 'waisted'**
   See also **1** in Guide to Bacchini (*p. 72*)
   ▶ *Neoascia* p. 200
   ▶ *Sphegina* p. 198

b) **Abdomen 'normal'**
   ▶ **2**

Sphegina sp.

2
from
1b

a) **Abdomen distinctly metallic**

Beware – the soldierfly *Chloromyia formosa* may be mistaken for *Lejogaster*. Check the wing venation to make sure it is a hoverfly.

▶ *Lejogaster* p. 210

b) **Abdomen with metallic or shining margin distinct from dull upper surface**
   ▶ **3**

c) **Abdomen non-metallic**
   ▶ **5**

CHRYSOGASTRINI

a) **Thorax metallic bronzy; abdomen segment edges: T1–T4 shiny**
See accounts for differentiating features.
▶ *Orthonevra* p. 208
▶ *Riponnensia* p. 208

b) **Thorax dark, often matt or with greenish/purplish reflections; abdomen segment edges: T1 dull, T2–T4 shiny**
▶ **4**

a) **Antennae black**
▶ *Melanogaster* p. 204

b) **Antennae orange to brown**

Beware – some females can have rather darkened antennae.

▶ *Chrysogaster* p. 206

a) **Abdomen with yellow markings**
▶ *Myolepta* p. 217

Beware – the yellow markings may not always be obvious, especially when the wings are closed over the back.

b) **Abdomen orange-brown and thorax either orange-brown or grey**
May darken to deep red-brown with age.
Thorax orange-brown
▶ *Hammerschmidtia* p. 216

Thorax grey
▶ *Brachyopa* p. 212

Brachyopa sp.

Sphegina

from 1a p.196

4 British and Irish species (3 illustrated)

Small, delicate, wasp-waisted hoverflies that fly with legs dangling. Usually encountered in dappled light around woodland umbellifers. The eyes are separated in both sexes. Only likely to be confused with *Neoascia* (p. 200) or *Baccha* (p. 74). More abundant in the north and west, probably because they favour damp habitats. The larvae live in decaying sap under bark, usually in damp situations such as logs lying on wet ground or partly submerged in pools or streams.

Beware – all four *Sphegina* need careful examination to differentiate (see table *opposite*). The dark and mixed colour forms of *S. sibirica* (p. 200) can be particularly confusing and require a check of the sides of the thorax (lower thoracic plurae) and the underside of abdomen segment S1.

LC *Sphegina elegans*

GB: Frequent =
Ir: Rare

Wing length: 5·0–6·75 mm

Identification: Like *Sphegina clunipes* and *S. verecunda* (N/I), *S. elegans* has a chitinous plate forming the underside of abdomen segment S1 (see *opposite*). It has yellowish humeri that usually contrast with the otherwise dark thorax. This character alone should ensure confident identification, but in males the genitalia are far smaller than those of *S. clunipes*.

Similar species: Most likely to be confused with dark and mixed colour forms of *Sphegina sibirica* (p. 200), but that species has a longer abdomen and the underside of abdomen segment S1 lacks a chitinous plate.

Observation tips: Found mainly in wet deciduous woodlands where it visits umbellifer flowers. Far less common than either *S. clunipes* or *S. sibirica* across most of Britain, and known from fewer than 20 records from south-west Ireland.

J F M A M J J A S O N D

LC *Sphegina clunipes*

GB: Widespread =
Ir: Frequent

Wing length: 4·75–7·0 mm

Identification: One of three very similar *Sphegina* species that have a chitinous plate forming the underside of abdomen segment S1 (see *opposite*). *S. clunipes* has black humeri and the R–M cross-vein lies well beyond the point where the sub-costa meets the costa. Males have very long and distinctive genital processes.

Similar species: Confused mainly with *Sphegina verecunda* (no species account, but see *opposite*), which differs in its male genitalia and the R–M cross-vein that lies ± opposite the point where the sub-costa meets the costa.

Observation tips: Usually found around and underneath white umbellifers in damp shaded places; it also visits Tormentil flowers on open ground around woodland in more northerly locations. Most common across western Britain and especially Scotland; it is widespread in Ireland but seemingly absent from the north-east.

J F M A M J J A S O N D

Sphegina species not otherwise covered:

S. verecunda	The scarcest *Sphegina* and generally found in wetter areas of woodland.

CHRYSOGASTRINI: *Sphegina*

Identification pointers for *Sphegina* species.

SPECIES	ABDOMEN UNDERSIDE S1	THORAX	♂ GENITALIA	♀ WING R–M CROSS-VEIN
S. elegans (*opposite*)	chitinous plate **present**	HUMERI **yellowish**		
S. clunipes (*opposite*)		HUMERI **black**	♂ GENITALIA **long, pointed process**	lies **well beyond** the point where the sub-costa meets the costa
S. verecunda (N/I)			♂ GENITALIA **rounded**	lies **± opposite** the point where the sub-costa meets the costa
S. sibirica (p. 200)	chitinous plate **absent**	shiny area above hind coxa		

abdomen underside base

S. sibirica (p. 200)

membranous, no chitinous plate

clunipes, S. elegans, S. verecunda

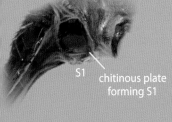

S1 — chitinous plate forming S1

♂ genitalia

S. clunipes long, pointed processes

S. verecunda rounded

♀ *Sphegina elegans* × 8

HUMERI **yellow**

♀ *Sphegina clunipes* × 8

HUMERI **black**

LC *Sphegina sibirica*

GB: Frequent =
Ir: Rare

Wing length: 7 mm

Identification: Slightly larger than *Sphegina clunipes* (p. 198), with shining patches on the sides of the thorax (lower thoracic plurae) and black 'feet' contrasting with the pale tibia. The underside of the base of the abdomen is entirely membranous with no chitinous plate forming underside of abdomen segment S1. Very variable in colour from all-dark to all-yellow, with some individuals having yellow-and-black piebald patterns.

Similar species: Yellow forms are unmistakable. Partially dark forms may look like *Sphegina elegans* (p. 198). Completely dark forms are difficult to separate from other dark *Sphegina* without checking the lower thoracic plurae and the underside of abdomen segment S1 for the presence/absence of a chitinous plate.

Observation tips: Mainly a northern and western species. It is a relatively recent colonist that is found mostly in coniferous woodland. Its powers of dispersal are quite remarkable, having been found on offshore islands and well above the tree-line on Scottish mountains. It is known from just four Irish records to date, but seems likely to greatly expand its range in due course.

J F M A M J J A S O N D

---

from 1a
p. 196

Neoascia

6 British and Irish species (4 illustrated)

Tiny black hoverflies. The females, especially, are wasp-waisted; males are less so but are somewhat elongate and often have a pronounced genital capsule. The eyes are separated in both sexes. Only likely to be confused with *Sphegina* (p. 198) or *Baccha* (p. 74). Larvae have been found in wet manure, compost and decaying vegetation around ponds/ditches.

LC *Neoascia meticulosa / tenur*

GB: Frequent/Widespread ▽
Ir: Local/Frequent

Wing length: 3·0–5·5 mm

Identification: The outer cross-veins of the wing are clear, and abdominal markings are absent or indistinct. The chitinous bridge behind the hind coxae (see *opposite*) is incomplete in *N. meticulosa* but complete in *N. tenur*; also the apex of the hind femur is yellow (can be indistinct) in *N. meticulosa*, black in *N. tenur* (see *opposite*).

Similar species: *Neoascia geniculata* (N/I) looks very similar to *N. tenur* but the chitinous bridge is incomplete (complete in *N. tenur*). These species are difficult to separate from photographs, although occasionally the yellow tip of the hind femur of *N. meticulosa* is sufficiently well depicted to make a confident identification. In theory, *N. tenur* and *N. geniculata* can be separated on the shape of the 3rd antennal segment, but this character is difficult to assess and is often misinterpreted, hence the need to check the chitinous bridge.

J F M A M J J A S O N D

Observation tips: Adults occur in wetlands and often occur together, with *N. tenur* perhaps occurring more frequently in acidic wetlands. The map shows their combined distribution in both Britain and Ireland, which suggests that they are primarily northern and western species.

CHRYSOGASTRINI: *Sphegina* | *Neoascia*

♀ *Sphegina sibirica* × 6

underside of abdomen base membranous, with no chitinous plate

Identification pointers for *Neoascia* species

SPECIES	N. tenur (opposite)	N. meticulosa (opposite)	N. geniculata (N/I)	N. obliqua (N/I)	N. interrupta (p. 202)	N. podagrica (p. 202)
WINGS	CROSS-VEINS **clear**			CROSS-VEINS **clouded**		
HIND COXAE	CHITINOUS BRIDGE **complete**	CHITINOUS BRIDGE **incomplete**		CHITINOUS BRIDGE **incomplete**		CHITINOUS BRIDGE **complete**
OTHER	HIND FEMORA **black** at apex	HIND FEMORA **yellow** at apex	HIND FEMORA **black** at apex	ABDOMEN T4 plain	ABDOMEN T4 with markings	ABDOMEN T4 plain

Neoascia chitinous bridges

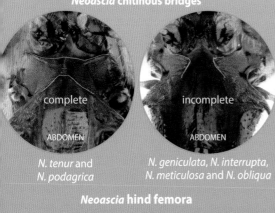

complete
ABDOMEN
N. tenur and *N. podagrica*

incomplete
ABDOMEN
N. geniculata, N. interrupta, N. meticulosa and *N. obliqua*

♀ *Neoascia meticulosa* × 10

Neoascia hind femora

N. tenur **black** at apex

N. meticulosa **yellow** at apex

Neoascia podagrica

LC / **LC**

GB: Widespread ▽
Ir: Widespread

Wing length: 3·5–5·0 mm

Identification: There are three *Neoascia* species in which there is clouding over the outer cross-veins, but this is the only one in which the chitinous bridge is complete (visible from certain angles examined from the rear and underneath – see *opposite*). There are markings on abdomen segments T2 and T3, which are often difficult to discern in resting individuals.

Similar species: *Neoascia obliqua* (N/I) and *N. interrupta*, which is unique within the genus in having spots on the side of abdomen segment T4 (although these markings can be obscure).

Observation tips: A widespread and common species in a range of habitats such as woodland rides and hedgerows, and around ponds and ditches. It is often found visiting low-growing flowers and has a long flight season.

J F M A M J J A S O N D

Neoascia interrupta

LC / **LC**

Nationally Scarce
GB: Local =
Ir: Not recorded

Wing length: 3·5–5·0 mm

Identification: The outer cross-veins of the wing are clouded and abdomen segments T2–T4 have distinct yellow markings. This is the only British species that typically has distinct markings on T4 (although these can be reduced or obscure).

Similar species: Both *Neoascia obliqua* and *N. podagrica* have clouds over the outer cross-veins but are much darker and less well-marked overall. Beware, male *N. interrupta* with reduced markings on T4 reduced are easily misidentified.

Observation tips: This is mainly a wetland species that is predominantly coastal and southern in distribution, although it does occur inland. Some association with Bulrush and with Hoary Willowherb is suspected but the evidence is circumstantial.

J F M A M J J A S O N D

***Neoascia* species not otherwise covered:**

N. geniculata	Of the three *Neoascia* species with clear wings, this is the most difficult to recognise. Confusion with *N. tenur* is most likely as both species have wholly black hind femora (variably yellow-tipped in *N. meticulosa*). Antennal shape is used by Stubbs & Falk (2002) to separate *N. tenur* and *N. geniculata*, but this is open to misinterpretation; therefore it is best to check the chitinous bridge behind the hind coxae (complete in *N. tenur*, incomplete in *N. geniculata*). The most readily visible identification characters of this species are highly subjective, and misidentifications are extremely common. It is therefore necessary to check against reliable voucher specimens to arrive at an accurate determination.
N. obliqua	One of the three *Neoascia* species with clouded outer cross-veins. It is a widely distributed but scarce species that is normally associated with beds of Butterbur.

CHRYSOGASTRINI: *Neoascia*

Neoascia podagrica × 10

WING cross-veins clouded

HIND COXAE chitinous bridge complete

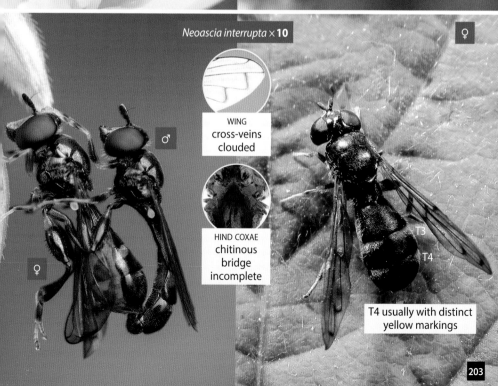

Neoascia interrupta × 10

WING cross-veins clouded

HIND COXAE chitinous bridge incomplete

T3
T4

T4 usually with distinct yellow markings

METALLIC CHRYSOGASTRINI

Differentiating features of partially metallic and metallic Chrysogastrini (see *opposite*)

GENUS	THORAX AND ABDOMEN	OTHER FEATURES
Melanogaster (*opposite*)	THORAX dark, often matt or with greenish/purplish reflections	ANTENNA **black**
Chrysogaster (*p. 206*)	ABDOMEN flattened with a matt centre contrasting with shiny edges to **T2–T4**, with **T1 dull**	ANTENNA **brown/orange**, although some may be darker
Orthonevra (*p. 208*)	THORAX bronzy metallic ABDOMEN flattened with a matt centre contrasting with shiny edges to **T1–T4**	FACE dust bands beneath antennae **thin** – see *page 209*
Riponnensia (*p. 208*)		FACE dust bands beneath antennae **broad** – see *page 209*
Lejogaster (*p. 210*)	ABDOMEN **wholly metallic**	

from 4a
p. 197

Melanogaster   2 British and Irish species (1 illustrated)

These are small black hoverflies in which the female does not have a 'nose' and that of the male is poorly developed. The abdomen is rather flattened with a matt centre contrasting with shiny edges (most noticeable in females). Identification can be difficult, requiring good magnification and lighting to separate the two species. They occur in wet meadows and along flowery heathland edge. The larvae live among emergent plants at the edges of ponds and ditches.

Melanogaster hirtella

GB: Widespread ▽
Ir: Frequent

Wing length: 5–6 mm

Identification: Females are more straightforward to identify than males. Their facial profile has no indication of a 'nose', their antennae are completely black, their abdomen is matt black with shiny edges (segments T2–T4 only), and there are upright yellowish hairs on the thorax. Males are more difficult to identify because they do have a slight 'nose' and are easily mistaken for small *Cheilosia* (*pp. 170–191*), particularly *C. antiqua* (see *Similar species below*).

Similar species: *Melanogaster aerosa* (N/I) is very difficult to separate and requires comparative specimens to check subtle differences in hair colours. Male *Cheilosia antiqua* (N/I, but see *p. 174*) are also similar in having completely black legs and hairless eyes but can be differentiated by the presence of a zygoma.

Observation tips: An abundant spring species, occurring in damp meadows where it visits buttercups and other flowers.

J F M A M J J A S O N D

Melanogaster species not otherwise covered:

M. aerosa	Mainly northern and western in acidic wetlands, but rarely abundant. Separation from *M. hirtella* is tricky (see account (*above*) for differences).

CHRYSOGASTRINI: *Melanogaster*

abdomens of the partially metallic and metallic Chrysogastrini

Lejogaster   *Riponnensia/ Orthonevra*   *Chrysogaster/ Melanogaster*

wholly metallic   top dull; segments T1–T4 with metallic edges   top dull; segments **T1 dull**; T2–T4 with metallic edges

Melanogaster hirtella × **10**

♀   ♂

♂ *M. hirtella* has a slight 'nose'.

Chrysogaster

3 British and Irish species (2 illustrated)

from 4b p.197

Small, shiny black hoverflies with somewhat metallic reflections. Closely related to *Lejogaster* (p. 210), *Melanogaster* (p. 204), *Orthonevra* and *Riponnensia* (both p. 208), which are all broadly similar in general appearance. Confusion is most likely with *Cheilosia* (pp. 170–191), but these are not metallic and have a more pronounced facial prominence. They are flower visitors whose larvae are aquatic, living in mud and among decaying vegetation at the edges of water bodies.

Chrysogaster cemiteriorum

GB: Frequent ▽
Ir: Local

Wing length: 5·25–6·5 mm

Identification: The combination of a slightly metallic body and strongly yellow wing bases makes this a fairly distinctive species. The side of the thorax above the front coxae is usually grey-dusted.

Similar species: Within this genus, most likely to be confused with *Chrysogaster virescens* (N/I), which has a more metallic green hue but is best separated by the lack of dusting on the sides of the thorax – but this is variable in *C. cemiteriorum* so care is needed. Confusion when at rest is possible with *Cheilosia impressa* (p. 190) and *Myolepta dubia* (p. 217): both have, or give the impression of having, strongly yellowed wing bases. From some angles, males can be tricky to separate from *Chrysogaster solstitialis* in photos.

Observation tips: A woodland species that is generally less abundant than *C. solstitialis*, with which it is often found. It visits a range of umbellifers, and although widely distributed across Britain it seems to have a far more restricted distribution in Ireland.

J F M A M J J A S O N D

Chrysogaster solstitialis

GB: Widespread =
Ir: Frequent

Wing length: 6·0–7·25 mm

Identification: A small and very dark-looking hoverfly with a matt black body with purplish reflections on the thorax (depending on the light angle) and strongly darkened wings. Males have strikingly bright-red eyes giving them a very distinctive appearance.

Similar species: Other *Chrysogaster* (see *above*); confusion is also possible with some housefly-like species in other families, which are rather black and have bright-red eyes.

Observation tips: A common species of woodlands, road verges and hedgerows, including areas that do not appear to be particularly wet. Adults visit a wide variety of umbellifers, especially Hogweed and Wild Angelica. Although widely distributed and often abundant in Britain, this species seemingly has a far patchier distribution in Ireland.

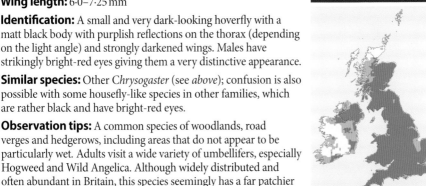

J F M A M J J A S O N D

***Chrysogaster* species not otherwise covered:**

C. virescens	A difficult species to identify and needs comparison with known specimens of *C. cemiteriorum* (see account (*above*) for differences). Widespread but very local in the north and west in acidic wetlands, but also occurs in bogs on heathland in southern England.

CHRYSOGASTRINI: *Chrysogaster*

See *page 204* for a summary of differentiating features of *Melanogaster/Chrysogaster*, *Lejogaster* and *Orthonevra/Riponnensia*, and *page 205* for a comparison of their abdomens.

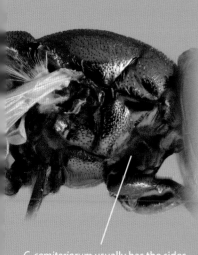

C. cemiteriorum usually has the sides of the thorax grey-dusted just above the front coxae.

♀ *Chrysogaster cemiteriorum* × **8**

♂   *Chrysogaster solstitialis* × **8**   ♀

> See *page 204* for a summary of differentiating features of *Melanogaster/Chrysogaster*, *Lejogaster* and *Orthonevra/Riponnensia* and *page 205* for a comparison of their abdomens.

from 3a
p. 197

Orthonevra
4 British and Irish species (1 illustrated)

Small, somewhat metallic hoverflies with a slightly flattened abdomen. Abdomen segment T1 has shiny, metallic edges (like *Riponnensia*), rather than the dull edges of *Melanogaster* (*p. 204*) and *Chrysogaster* (*p. 206*). *Lejogaster* (*p. 210*) has a completely shiny and less flattened abdomen. The presence of a horizontal, dusted band across the face below the antennae is a further useful character. The larvae live among emergent plants at the edges of ponds and ditches.

LC / **LC**
Orthonevra nobilis

GB: Frequent ▽
Ir: Rare

Wing length: 4·0–5·75 mm

Identification: A small, metallic hoverfly with black legs. Antenna segment 3 is about 2× as long as wide and rather pointed. There is a narrow dusted band across the face below the antennae. There is often a slight darkening in the centre of the wing.

Similar species: *Orthonevra geniculata* (N/I) and *O. intermedia* (N/I), which both have partially yellow legs; *O. brevicornis*, which has shorter and blunter antennae; *Riponnensia*, which has a broad dusted band below the antennae; and *Lejogaster* (*p. 210*), which have a fully metallic abdomen.

Observation tips: A wetland species which visits a wide range of flowers, especially umbellifers. However, it seems to disperse quite widely and is frequently seen at Wild Parsnip and Wild Carrot in drier grasslands. Widely distributed in Britain, and most frequent towards the south-west; there have been 21 records from Ireland.

J F M A M J J A S O N D

from 3a
p. 197

Riponnensia
1 British and Irish species (illustrated)

Formerly placed in *Orthonevra*, with which it shares many characters, such as the abdomen having a flattened matt upper surface and segment T1 having a shiny metallic edge. The larvae live in accumulations of wet, rotting vegetation at the edges of ponds and ditches.

LC / **LC**
Riponnensia splendens

GB: Widespread ▽
Ir: Frequent

Wing length: 5·5–7·0 mm

Identification: Larger than most *Orthonevra*, with a broad, dusted band beneath the antennae.

Similar species: *Orthonevra nobilis*, which has a narrow dusted band below the antennae; *Lejogaster* species (*p. 210*), which have a fully metallic abdomen; and the soldierfly *Chloromyia formosa* (see *page 211*) (check the wing venation to make sure it is a hoverfly).

Observation tips: A wetland species that occurs in marshy places around ponds, rivers and canals, and also in wet woodlands. It frequently visits umbellifers but can be seen at many other flowers, including Ivy in the autumn. In Britain, it is most frequent in the south. In Ireland, it seems to have a patchier distribution.

J F M A M J J A S O N D

CHRYSOGASTRINI: Orthonevra | Riponnensia

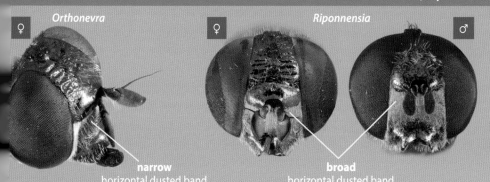

Orthonevra — narrow horizontal dusted band

Riponnensia — broad horizontal dusted band

♀ *Orthonevra nobilis* × 8

Orthonevra species not otherwise covered:

O. brevicornis	Scarce, but widely distributed in marshes and fens in England and Wales (very few Scottish records).
O. geniculata	Scarce, but widely scattered in fens and mildly acidic boggy areas. There are 49 records from Ireland, suggesting that it is widespread there.
O. intermedia	New for Britain when discovered at two sites in Cheshire in 2006.

Riponnensia splendens × 8

CHRYSOGASTRINI: *LEJOGASTER*   Differentiating features of metallic Chrysogastrini: *p. 204*

> See *page 204* for a summary of differentiating features of *Melanogaster/Chrysogaster*, *Lejogaster* and *Orthonevra/Riponnensia* and *page 205* for a comparison of their abdomens.

from
2a
p. 196

Lejogaster   2 British and Irish species (both illustrated)

Small, dark hoverflies with a metallic sheen. The eyes are separated in both sexes. There are few other hoverflies with which confusion is likely, the main one being *Riponnensia splendens* (p. 208). The larvae are aquatic and live in wet, decaying plant material in the edges of ponds and ditches.

LC
LC
Lejogaster metallina

GB: Widespread ▽
Ir: Frequent

Wing length: 4·75–6·5 mm

Identification: A small, metallic-green hoverfly which has comparatively large antennae. Males have completely black antennae. Females have relatively short and round antennae which vary in colour between individuals – from completely black to showing some orange/yellow on the lower part of segment 3.

Observation tips: A species of damp grasslands and wetlands that visits yellow flowers such as buttercups. Easily overlooked among other chrysogastrines (*p. 196*) and *Cheilosia* (*pp. 170–191*) and best found by sweeping. Widely distributed across Britain but much commoner in northern and western areas. In Ireland it is more abundant in the south with a more patchy distribution farther north.

Similar species: The two *Lejogaster* species need careful differentiation (see table *opposite*). They may be confused with *Orthonevra nobilis*, *O. intermedia* and *Riponnensia splendens* (all *p. 208*), but the abdomen of these three species has a matt upper surface with shiny edges, rather than being wholly metallic. Beware – the very common soldierfly *Chloromyia formosa* (inset *opposite*) may be mistaken for *Lejogaster* species. Check the wing venation to make sure it is a hoverfly.

J F M A M J J A S O N D

LC
LC
Lejogaster tarsata

GB: Local =
Ir: Scarce

Wing length: 4·5–6·0 mm

Identification: Smaller and more delicate than *Lejogaster metallina*, and tends to have more bluish reflections, especially on the thorax of the male and the abdomen of both sexes. The antennae of the male are rather roundish and partially yellow; those of the female are relatively elongated and always yellowish (usually extensively so) on the lower part.

Observation tips: Mostly coastal, especially on brackish coastal marshes but also some riverine wetlands. Sometimes abundant on Scottish coastal seepage marshes. Confined to a very few localities in Ireland.

J F M A M J J A S O N D

CHRYSOGASTRINI: *Lejogaster*

Tips for separating *Lejogaster* species

SPECIES	COLOUR AND FORM	♂ ANTENNA	♀ ANTENNA
L. metallina	metallic green; larger and broader than *L. tarsata*	**black**	variable; black to partially orange/yellow; SEG. 3 **shorter and rounder** than in *L. tarsata*
L. tarsata	metallic bluish-green; smaller and more slender than *L. metallina*	**partially yellow**	partially yellow; SEG. 3 **more elongate** than in *L. metallina*

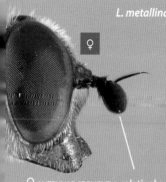

L. metallina

♂ *Lejogaster metallina* × 8

♀ ANTENNAE SEGMENT 3 relatively short and round; colour varies between individuals from partially orange/yellow on the lower part to completely black (as here).

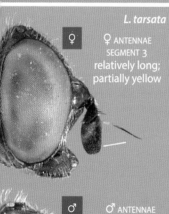

L. tarsata

♀ ANTENNAE SEGMENT 3 relatively long; partially yellow

♀ *Lejogaster tarsata* × 8

The soldierfly *Chloromyia formosa*

♂ ANTENNAE SEGMENT 3 rounded; partially yellow

Brachyopa

from 7c p.59 and 5b p.197

4 British and Irish species (all illustrated)

Orange-brown hoverflies that require careful examination under magnification to determine the presence and shape of a pit on inner face of the 3rd antennal segment. Rarely identifiable in the field (but sometimes possible from high-resolution photographs). They are often regarded as difficult to find and rare but are probably more widespread than records suggest. Adults hover close to sap runs or bask on tree trunks or sunlit leaves. The larvae feed on micro-organisms in fermenting sap. There might be confusion with the very rare *Hammerschmidtia ferruginea* (p.216) but that species is confined to a few locations in the highlands of Scotland. Confusion with some other fly families such as scathophagids, muscids and psilids is also possible, but they all lack a vena spuria and have copious bristles.

LC *Brachyopa insensilis*

GB: Local =
Ir: Rare

Wing length: 6·5–7·25 mm

Identification: The only British *Brachyopa* that lacks (or has only an indistinct) pit on the inside face of the antennae. The partially grey-dusted scutellum, yellowish humeri (although rather variable and often darker) and almost bare aristae are useful confirmatory features.

Similar species: All four *Brachyopa* require careful examination of the antennal pit and the colour of the scutellum and humeri.

Observation tips: Particularly frequent in urban areas where it is mostly found at Horse-chestnut sap runs. This association may point towards *B. insensilis* but should not be relied upon as other species may also be present. Widely distributed across Britain but seemingly confined to a few localities in Ireland, where more intensive searches of urban habitats may prove it to be more widespread.

J F M A M J J A S O N D

LC *Brachyopa scutellaris*

GB: Frequent ▲
Ir: Local

Wing length: 6·5–7·75 mm

Identification: The antennal pit is kidney-shaped, the aristae are pubescent and the humeri are usually yellowish (but can be dark).

Similar species: Confusion with *Brachyopa pilosa* (p. 214) is most likely, due to the humeri of *B. scutellaris* being individually variable in colour and the dusting on *B. pilosa* humeri not always being obvious. The shape of the antennal pits must be checked (difficult to see in the majority of photos and can be missed or distorted by angles of view). Dark individuals in particular can be mistaken for *B. bicolor* (p. 214) if this feature is misinterpreted.

Observation tips: Larvae live in sap runs low down on the trunks of deciduous trees. Males bask on sunlit leaves, especially those of Sycamore and limes, and hover around the sunlit bases of trees. The flight-period is short but, when found, this species can be surprisingly numerous. Widely distributed in woodlands across England and Wales but largely confined to the south-west of Scotland and Speyside. It is seemingly very rarely recorded in Ireland, although this may also reflect recording effort and a paucity of deciduous woodland.

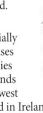

J F M A M J J A S O N D

CHRYSOGASTRINI: *Brachyopa*

Beware – *Brachyopa* can only be separated reliably by the presence/absence and shape of pits on the antennae. These are indistinct/absent in *B. insensilis*, large and kidney-shaped in *B. scutellaris* and small and rounded in *B. pilosa* and *B. bicolor*. Beware that the actual shape of the pits can be difficult to discern unless seen straight-on. This is a difficult character which normally needs microscopic examination and known specimens for comparison.

Brachyopa antennal pits

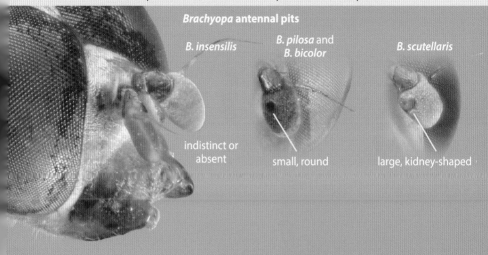

B. insensilis — indistinct or absent

B. pilosa and *B. bicolor* — small, round

B. scutellaris — large, kidney-shaped

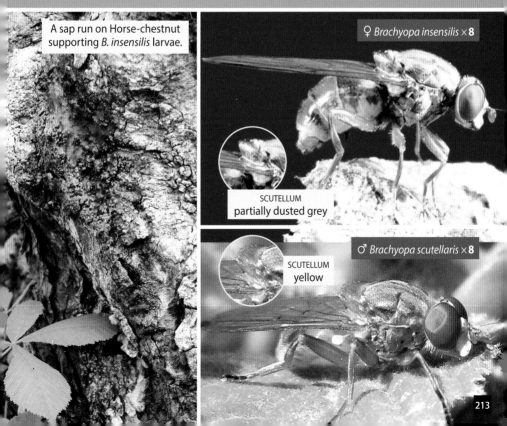

A sap run on Horse-chestnut supporting *B. insensilis* larvae.

♀ *Brachyopa insensilis* × 8
SCUTELLUM partially dusted grey

♂ *Brachyopa scutellaris* × 8
SCUTELLUM yellow

Brachyopa bicolor

Wing length: 6·5–9·25 mm

Identification: This species is normally somewhat larger than other *Brachyopa*, but size is very variable. In photographs, the grey-dusted and somewhat darker front part of the scutellum is clearly separated from the yellow part to the rear by a transverse central groove (see *opposite*). The antennae have a round pit (see *page 213*), but this is difficult to see in most photos.

Similar species: Other *Brachyopa*, although these lack the distinctive coloration and structure of the scutellum of *B. bicolor*. Dark individuals of *B. scutellaris* (*p. 212*) are sometimes confused when the shape of the antennal pits is misinterpreted.

Observation tips: Rarely seen far from the sap runs that its larvae utilise, although in the New Forest, Hampshire it can be abundant at the flowers of Rowan and Hawthorn. Finding this species at sap runs involves similar techniques to other *Brachyopa* – first find your sap run, or check out the sunlit bases of trees. Whilst it is much the rarest of the four British *Brachyopa* species, it has increased greatly in abundance in recent years and may be spreading northwards.

J F M A M J J A S O N D

Brachyopa pilosa

Wing length: 7·5–8·5 mm

Identification: The scutellum is undusted and the aristae are pubescent. There is a small round pit on the inner face of the antennae (see *page 213*) and the humeri are grey-dusted.

Similar species: Confusion with *Brachyopa scutellaris* (*p. 212*) is most likely, due to the colour of the humeri of that species being extremely variable and the dusting of *B. pilosa* not always being obvious. It is important to check the shape of the antennal pits (which are not readily visible in the majority of photos and can be missed or distorted by angles of view).

Observation tips: There are two separate populations of *B. pilosa*: one in southern England and the other in Scotland. They are occasionally found at flowers such as Rowan and Hawthorn, but are more likely to be seen visiting sap runs, especially those on Beech and Aspen. It has also been found on the intact bark of fallen Beech trees and on cut logs of Aspen.

J F M A M J J A S O N D

CHRYSOGASTRINI: *Brachyopa*

♂ *Brachyopa bicolor* × **8**

SCUTELLUM dark at front and yellow at the rear, with a distinct central groove

♂ *Brachyopa pilosa* × **8**

CHRYSOGASTRINI: *Hammerschmidtia*

from 7c
p. 59
and
5b
p. 197

Hammerschmidtia

1 British species (illustrated)

Unmistakable – once you realise that it is a hoverfly. In general shape it looks more like a fly from the families Psilidae or Muscidae. Larvae live in rotting sap under the bark of recently fallen, mature Aspen logs. These only provide suitable conditions for a limited period of time, perhaps two to three years. Woodlands with sufficient Aspen to ensure a continual supply of suitable fallen timber are now very scarce.

LC *Hammerschmidtia ferruginea*
EN Aspen Hoverfly

UKBAP Priority Species
GB: Rare [no trend data]
Ir: Not recorded

Wing length: 8·25–9·75 mm

Identification: An unmistakable uniformly red-brown hoverfly; young individuals are pale yellow-brown, becoming considerably darker, even dark brown, with age.

Similar species: *Brachyopa* (p. 212) have a similar orange-brown abdomen but the thorax is grey; however, those species do not darken with age to the same degree as *Hammerschmidtia ferruginea*. Some completely brown Psilidae might be mistaken for *H. ferruginea* in the highlands of Scotland (but occur much more widely); they lack outer cross-veins and a vera spuria.

Observation tips: An extremely rare hoverfly known from just a handful of Aspen woods in northern Scotland. Recent studies have shown that although the population is very restricted, a single log is capable of supporting a surprisingly large number of larvae. These studies also found that adults are capable of dispersing over several kilometres to find new breeding sites.

J F M A M J J A S O N D

Hammerschmidtia ferruginea × 5

CHRYSOGASTRINI: *Myolepta*

Myolepta

2 British species (1 illustrated)

Easily recognised by the broad yellow margins on abdomen segments T1, T2, and often T3, which create a central black stripe. The larvae live in rot-holes in various deciduous trees.

LC *Myolepta dubia*

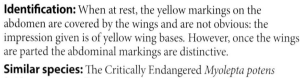

Nationally Scarce
GB: Local =
Ir: Not recorded

Wing length: 6·0–8·75 mm

Identification: When at rest, the yellow markings on the abdomen are covered by the wings and are not obvious: the impression given is of yellow wing bases. However, once the wings are parted the abdominal markings are distinctive.

Similar species: The Critically Endangered *Myolepta potens* (N/I) – separation from *M. dubia* can be difficult as features are subjective (♂ face; ♀ T2 markings) and needs expert confirmation. A *M. dubia* with closed wings could easily be dismissed as another species with actual yellow wing bases, such as a some what elongate *Cheilosia impressa* (p. 190) or *Chrysogaster cemiteriorum* (p. 206).

Observation tips: Mainly a southern species, occurring not only in woodlands but also in open countryside where there are old trees. Adults visit flowers, especially umbellifers. Searching for larvae in rot-holes, particularly in Horse-chestnut and Beech, may show it to be even more widespread. It appears to be becoming more common and may be expanding its range.

J F M A M J J A S O N D

Myolepta species not otherwise covered:

| *M. potens* | CR BAP | Once known solely from two sites in Somerset, it is now known from one new Somerset site, the Forest of Dean, Gloucestershire and Moccas Park NNR, Herefordshire. |

The abdomen markings of *Myolepta* are distinctive.

♂ *Myolepta dubia* × 5

ERISTALINI | MERODONTINI (*MERODON*)

from
2a, 2b
p. 54

Guide to Eristalini (and Merodontini: *Merodon*)

The Eristalini are distinctive as their wings have a prominent loop in vein R_{4+5}. Once this feature is learned, the tribe is instantly recognisable. However, this loop also occurs in *Merodon equestris* (Merodontini (p. 248)), hence it is included here. There are eight genera which form two distinct groups: those with distinct longitudinal stripes on the thorax (*Anasimyia*, *Helophilus*, *Lejops* and *Parhelophilus*) and those that are at least vaguely honey-bee mimics (*Eristalis*, *Eristalinus*, *Mallota* and *Myathropa*).

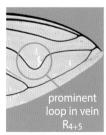

prominent loop in vein R_{4+5}

1
a) Bumblebee or honey-bee mimic
▶ 2 See also *Identifying wasp and bee mimics* pp. 62–65

b) Not a bee mimic
▶ 3

2
from 1a

a) **Bumblebee mimic:**
   Scutellum ground colour black
   Hind femur greatly enlarged and with a triangular projection below
   Hind legs entirely black
   ▶ *Merodon equestris* p. 248

b) **Bumblebee mimic:**
   Scutellum ground colour yellow
   Hind femur normal
   Hind tibia partly pale
   ▶ *Eristalis intricaria* p. 228

 ♂

Beware – females and males look very different – a few individuals can look more like 'typical' *Eristalis*.

c) **Honey-bee mimic:**
   Large; hind legs modified
   – Hind tibia enlarged
   ▶ *Eristalis tenax* p. 224
   – Hind femur enlarged
   ▶ *Mallota cimbiciformis* p. 236

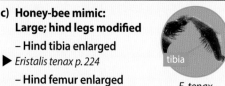
tibia
E. tenax
E. tenax

d) **Poor honey-bee mimics:**
   Hind legs not modified
   ▶ Most *Eristalis* pp. 220–233

ERISTALINI | MERODONTINI: *Merodon*

3
from
1b

a) **Eyes greenish and with reddish spots**
   Scutellum black
   ▶ *Eristalinus* p. 234

Beware – eye colour and most spotting is lost in preserved specimens.

b) **Thorax with pale bar running across creating a 'Batman' marking towards the back**
   ▶ *Myathropa florea* p. 236

Beware – these markings are very variable and can be obscure in some spring specimens, which tend to be darker and less colourful.

c) **Thorax with pale stripes running lengthways**
   ▶ **4**

4
from
3c

a) **Antennae black**

b) **Antennae orange**

HIND TIBIA:
one dark ring
▶ *Helophilus* pp. 242–245

HIND TIBIA:
one dark ring
▶ *Parhelophilus* p. 238

HIND TIBIA:
two dark rings
▶ *Lejops* p. 244

HIND TIBIA:
two dark rings
▶ *Anasimyia* p. 240

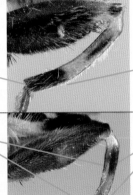

Eristalis

from 2a–c p. 218

10 British and Irish species (all illustrated)

This is one of the group of genera that have a loop in vein R_{4+5}. Together with *Eristalinus*, they have a 'petiole' beyond the junction of veins R_1 and R_{2+3} and no eye spots. Although often described as bee mimics, apart from *E. tenax* (which is a good honey-bee mimic) and the polymorphic bumblebee mimic *E. intricaria*, they are not very convincing. The larvae are aquatic 'rat-tailed maggots', that live in wet, decaying vegetation, usually in ponds or ditches, but can be found in farmyard manure pits and silage.

Mistaken identifications involving this genus dominate the errors in photographic recording. *Epistrophe eligans (p. 136)* is also mistaken, with some frequency, for several *Eristalis* species, but lacks a wing loop. These problems almost certainly arise because observers try to match abdominal markings rather than using the critical, and sometimes subtle, characters that differentiate the species.

Identifying *Eristalis*
These following pages show a step-by-step illustrated guide to the identification of *Eristalis* species. There are some subtle and often not-so-subtle differences in the extent and definition of characters, making it important to look at a combination of features. **For most *Eristalis*, the chances of an accurate identification from photographs is greatly improved if a combination of images from above, the side and the front can be taken.**

The key contains notes highlighting these potential pitfalls and most likely confusion species where appropriate. A recent key in *British Wildlife* (Ball & Morris 2022) provided detailed guidance on how to tackle this genus. It contains additional notes that may be helpful.

1 Overall form

from 1b p. 218

a) Bumblebee mimic:

Black hairs on sides of thorax below wing (from side)

▶ *Eristalis intricaria* p. 228

Beware – females and males look very different. Typically more furry than other *Eristalis* but a few well-marked individuals can be confusing. Check thorax hair colour to be sure. Confusion is more likely with other bumblebee mimics – *pp. 62–65*.

b) Honey-bee mimics (most are poor mimics):

Pale hairs on sides of thorax below wing (from side)

Honey-bee mimics (most are poor)

a) Front + middle tarsi wholly orange/yellow:

▶ *Eristalis pertinax* p. 224

NOTE: Usually an obvious species, but in case of doubt check the stigma, which is elongate.

b) Some front + middle tarsi partially darkened:

▶

Beware – variable; a few look all-pale, but typically some contrast between tarsal segs. 1 and 5.

Some front + middle tarsi partially darkened

a) Hind tibia entirely dark, slightly curved and distinctly enlarged:

▶ *Eristalis tenax* p. 224

NOTE: Also has a **dark band of hairs on the eyes**.

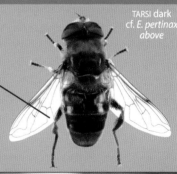

TARSI dark
cf. *E. pertinax*
above

b) Hind tibia entirely orange:

▶ *Eristalis cryptarum* p. 228
[A rare species restricted to a small area on south Dartmoor.]

Beware – can resemble ♂ *E. intricaria* at distance.

c) Hind tibia partially pale:

▶

ERISTALINI: *ERISTALIS*

4 Hind tibia partially pale

from 3c

a) Hind femora heavily grey dusted:

▶ *Eristalis similis* p.226
Occasional unlikely vagrant

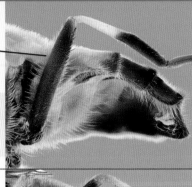

Beware – confusion is possible with *E. nemorum* especially. The stigma of *E. similis* is comparatively elongate.

b) Hind femora shiny:

▶ 5

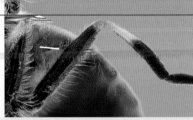

5 Hind femora shiny

from 4b

a) Hind metarsus pale:

▶ *E. rupium* p.230
Northern and western range

NOTE: Strong dark wing clouds

Beware – from some angles the downward-pressed hairs on the hind tarsi of *E. horticola* can look deceptively pale.

b) Hind metarsus dark:

▶ 6

6 Hind metatarsus dark

from 5b

a) Wing stigma sharply defined ▶ 7

---

a) Wing stigma diffuse ▶ 8

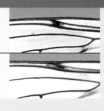

Beware – a difficult feature to assess; typically best used in conjunction with other features.

ERISTALINI: *Eristalis*

Wing stigma sharply defined

NOTE: The stigma is tricky to assess and is best used in conjunction with a view of the face, which has a clearly defined central undusted stripe.

▶ *Eristalis nemorum* p. 226

Beware – readily confused with *E. arbustorum*, which typically has a wholly dusted face, and *E. abusiva* which has a narrow central stripe.

Wing stigma diffuse

a) **Middle tibia at most slightly darkened at the tip (side view only):**

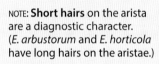

▶ *Eristalis abusiva* p. 232

NOTE: **Short hairs** on the arista are a diagnostic character. (*E. arbustorum* and *E. horticola* have long hairs on the aristae.)

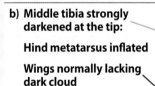

b) **Middle tibia strongly darkened at the tip:**

  Hind metatarsus inflated

  Wings normally lacking dark cloud

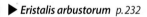

▶ *Eristalis arbustorum* p. 232

Beware – some females can have dark wing clouds.

c) **Middle tibia strongly darkened at the tip:**

  Hind metatarsus at most as thick as the tip of the tibia

  Wings normally with well-defined dark cloud

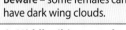
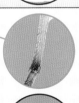
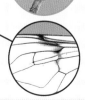
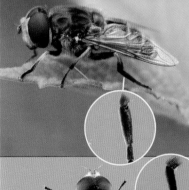

▶ *Eristalis horticola* p. 230

Beware – strong dark wing clouds occur in *E. rupium*; some *E. pertinax* and *E. arbustorum* can also show dark clouding.

ERISTALINI: *ERISTALIS*   See also *Identifying wasp and bee mimics pp. 62–65* | Guide to *Eristalis*: *p. 220*

Larger *Eristalis*   1/2

LC *Eristalis pertinax*

LC Wing length: 8·25–12·75 mm

Identification: One of the easiest *Eristalis* to identify, as the tarsi on the front and middle legs are usually orange/yellow (sometimes brownish). The male abdomen is markedly elongate and triangular, which conveys a distinctive jizz.

Similar species: Most frequently confused with *Eristalis tenax*, which has dark tarsi. Occasionally confused with *E. nemorum*, (*p. 226*) especially in photographs where the tarsi are obscured or partially pale in the latter species.

Observation tips: This species occurs almost everywhere, including upland moorland, throughout the warmer months (March to November). It is one of the first species to appear in the spring and males characteristically defend territories in woodland rides and around flowering bushes. Remarkably, very few females are seen at this time. It becomes even more abundant in late summer and can be seen in considerable numbers on the flowers of Wild Angelica, thistles and Ivy until the first frosts.

J F M A M J J A S O N D

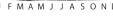

WING stigma **distinctly elongate**

LC *Eristalis tenax* Drone Fly

LC Wing length: 9·75–13·0 mm

Identification: Three characters distinguish this large honey-bee mimic from other *Eristalis*: the eyes have a vertical stripe of longer, dark hairs; the black facial stripe is very broad; and the hind tibia is distinctly enlarged, curved and entirely dark (partly pale in other *Eristalis*). The tarsi on the front and middle legs are dark.

Similar species: Most frequently confused with *Eristalis pertinax*, which has yellow tarsi. Also often misidentified as *E. arbustorum* (*p. 232*) on account of the stigma-shape, but that species has a partially yellow hind tibia. Could possibly be confused with *E. similis* (*p. 226*) but that species is heavily grey-dusted on the sides of the thorax and the hind femora and has a slightly differently shaped and more elongated stigma. Confusion with *Mallota cimbiciformis* (*p. 236*) is possible but in that species the hind femora are strongly thickened

J F M A M J J A S O N D

Observation tips: Occurs in the spring but most abundant in late summer and autumn, when it is often numerous on Ivy flowers. Females hibernate in sheltered cavities in caves and buildings. The larvae live in highly enriched aquatic environments, including slurry tanks and silage clamps, where they can sometimes occur in vast numbers.

WING stigma **diffuse and extended**

ERISTALINI: *Eristalis*

♂ *Eristalis pertinax* × 6

FACE narrow shining stripe

FRONT + MIDDLE TARSI yellow

♀ *Eristalis tenax* × 6

FACE very broad stripe

EYES vertical stripes of dark hairs

HIND TIBIA curved, enlarged, dark

Larger Eristalis 2/2

LC Eristalis similis
NE

Wing length: 10–11 mm

GB: Local [no trend data]
Ir: Not recorded

Identification: One of the larger *Eristalis* (but males are smaller than females). The sides of the thorax and the hind femora are grey-dusted; the stigma is somewhat elongate, the tarsi are dark and the hind tibia is strongly bicoloured (pale upper/dark below).

Similar species: As the name suggests, readily confused with other *Eristalis* – especially *E. pertinax* (*p. 224*), in which the stigma is similar but the front tarsi are yellow; *E. tenax* (*p. 224*), in which the hind tibia is somewhat curved, distinctly enlarged and entirely dark; and *E. nemorum*, which also has quite ashy thoracic pleurae but typically has a short quadrate stigma.

Observation tips: *E. similis* is a migrant that is becoming more frequent, but with numbers varying from year to year. It seems to be recorded most regularly by photographers, perhaps because they observe and photograph more *Eristalis* than most other hoverflies.

J F M A M J J A S O N D

WING stigma **somewhat elongate**

Smaller Eristalis 1/4

LC Eristalis nemorum
LC

Wing length: 8·25–10·5 mm

GB: Widespread =
Ir: Frequent

Identification: The most obvious identification feature is the behaviour of the male, which will hover above a female feeding on a flower in a pose that is much photographed (see *opposite* and *page 14*). Both sexes are variable in both size and coloration and identification requires care. The wing stigma is usually discrete and sharp-edged. The tarsi on the front and middle legs are at least partially dark but the extent varies considerably between individuals and they can appear yellow unless inspected closely.

Similar species: Confusion most likely with *E. arbustorum* or *E. abusiva* (both *p. 232*) when viewed from above, as judging stigma-shape is not always straightforward – in those species the stigma is more diffuse. *E. similis*, has grey-dusted hind femora and a somewhat elongated stigma. Occasionally confused with *E. pertinax* (*p. 224*), which has both front and middle tarsi wholly yellow.

Observation tips: Widespread, occurring in woodland rides, hedgerows and flowery meadows, often some distance from obvious breeding sites (shallow, nutrient-enriched water). Adults visit a wide variety of flowers. It has a long flight-period and can be abundant in midsummer.

J F M A M J J A S O N D

WING stigma usually **discrete and sharp-edged**

ERISTALINI: *Eristalis*

♀ *Eristalis similis* ×6

HIND FEMUR dusted

FACE narrow undusted stripe

THORAX SIDE grey-dusted

♂ *Eristalis nemorum* ×6

HIND LEG metatarsus dark; no thicker than tibia and not inflated as in *E. arbustorum* (see panel on *page 231*)

FACE well-defined undusted stripe

Characteristic behaviour in which males hover above the female.

Smaller *Eristalis* 2/4

Eristalis intricaria

GB: Widespread =
Ir: Widespread

Wing length: 8·25–12·0 mm

Identification: A sexually dimorphic bumblebee mimic that is furrier than other *Eristalis*, with a yellow scutellum. Males are dark with a reddish-brown 'tail', while females are somewhat larger with a white 'tail'. Unlike other bumblebee mimics, the tibiae are half black and half pale. Some males can have extensively orange abdomen segments T2 and T3 and look like other *Eristalis*, but they can be recognised by the black hairs on the thorax sides beneath the wings (other *Eristalis* have pale hairs in the same location).

Similar species: Most likely to be confused with *Cheilosia illustrata* (p. 176), but that species lacks a loop in wing vein R$_{4+5}$ and has a black scutellum. *Merodon equestris* (see pp. 29 and 248), has a black scutellum and all-black tibiae. *Chlorhinia ranunculi* (p. 292) has strongly swollen hind femora but lacks the loop in wing vein R$_{4+5}$. Confusion with *Eriozona syrphoides* (p. 116), *Leucozona lucorum* (p. 116) and *Volucella pellucens* (p. 272) is also possible, but those species similarly lack the wing loop characteristic of *Eristalis*.

J F M A M J J A S O N D

Observation tips: Usually found in or near damp places. Males characteristically hover close to nectar sources, such as willows, in the spring. Both sexes visit a range of low-growing flowers, with a liking for blue and purple ones such as thistles. There are two main peaks in abundance: a small one in the spring and a much larger one in midsummer.

Eristalis cryptarum

UKBAP Priority Species
GB: Rare [no trend data]
Ir: Rare

Wing length: 6·5–10·0 mm

Identification: This is the only *Eristalis* species with a completely orange hind tibia.

Similar species: Live *Eristalis cryptarum* appear to be smaller and dumpier than other *Eristalis* species, perhaps resembling male *E. intricaria* at a distance. *Sericomyia lappona* (p. 268) has also been misidentified as *E. cryptarum*.

Observation tips: This is an exceptionally rare species that is now confined in Britain to a few rhôs pasture sites on the south side of Dartmoor, Devon. There are, however, three records from south-west Ireland. It is a regular visitor to flowers of Bramble and Devil's-bit Scabious.

J F M A M J J A S O N D

Bog Hoverfly | A recording issue is that the colloquial name Bog Hoverfly is used by different guides for both *E. cryptarum* and *Sericomyia silentis* (p. 268). Consequently, most records of *E. cryptarum* on iRecord are erroneous (hence use of this name is not encouraged here).

ERISTALINI: *Eristalis*

Eristalis intricaria × 6

Eristalis cryptarum × 6

Smaller *Eristalis* 3/4

LC *Eristalis horticola*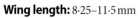

GB: Widespread ▽
Ir: Widespread

Wing length: 8·25–11·5 mm

Identification: Usually relatively large and more brightly marked than other *Eristalis*, with a prominent dark mark across the centre of each wing (although this is individually very variable in density and extent). The hind metatarsus is dark.

Similar species: Most likely to be confused with *Eristalis rupium* (*p. 230*), which has a pale hind metatarsus. There are several similar *Eristalis* in Europe that could potentially turn up in Britain, the separation of which is based on features of the male genitalia.

Observation tips: Mainly a woodland and hedgerow species, which is widespread but tends to be more abundant in the north. It is frequently found in association with Bramble and Hawthorn flowers. Males often hover and defend a territory in a similar manner to *E. pertinax* (*p. 224*).

J F M A M J J A S O N D

WING variable **dark mark across centre of wing**

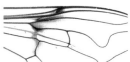

LC *Eristalis rupium*

GB: Frequent ▽
Ir: Not recorded

Wing length: 7·75–11·25 mm

Identification: The wings have an almost milky appearance with a dark chocolate marking that is more pronounced than in any other *Eristalis* (beware, however, that newly emerged individuals lack this marking). The hind metatarsus and the joint with tarsal segment 2 are pale, a character not exhibited by any other *Eristalis*.

Similar species: *Eristalis horticola* (*p. 230*), which has a dark hind metatarsus. There are several similar *Eristalis* in Europe that could potentially turn up in Britain, the separation of which is based on features of the male genitalia.

Observation tips: Mainly an upland species of northern and western Britain. However, in northern Scotland it occurs at low altitudes and has been noted in numbers on Hemlock Water-dropwort flowers along the shore of a sea-loch. Generally uncommon, but can be abundant where it occurs.

J F M A M J J A S O N D

WING extensive **dark marking highly conspicuous in mature individuals**

ERISTALINI: *Eristalis*

♀ *Eristalis horticola* × **6**

FACE broad undusted stripe

Hind metatarsi of *Eristalis rupium*, *E. horticola* and *E. arbustorum* compared

PALE
E. rupium
TIBIA

DARK;
at most as thick as the tip of the tibia
E. horticola
TIBIA

DARK;
inflated
E. arbustorum
(p. 232)
TIBIA

♀ *Eristalis rupium* × **6**

FACE thinly dusted, dark; narrow undusted stripe

TIBIA
METATARSUS
1
2
3
TARSUS

HIND METATARSUS
pale; thinner than tibia

Smaller Eristalis

LC Eristalis arbustorum

GB: Widespread =
Ir: Widespread

Wing length: 7–10 mm

Identification: A small *Eristalis* with an entirely yellow-dusted face that lacks a central shining black stripe. As an individual ages, the dusting on the face can become rubbed, allowing black patches to show through – so care is needed. The middle tibiae are pale and strongly darkened towards the metatarsal joint and the hind metatarsus is distinctly inflated. The stigma is diffuse and extended.

Similar species: Other small *Eristalis*, particularly *E. nemorum* (p. 226), which has a sharp-edged stigma, and the scarcer *E. abusiva* (see *opposite*). Confusion is also possible with *Epistrophe eligans* (p. 136) but that species lacks a loop in wing vein R$_{4+5}$.

Observation tips: Widespread and common. Adults occur a long way from breeding sites and visit a wide range of flowers such as knapweeds, ragworts, thistles and umbellifers. Numbers increase through the spring and it is most abundant in midsummer.

J F M A M J J A S O N D

WING Stigma diffuse and extended

LC Eristalis abusiva

GB: Frequent ▽
Ir: Frequent

Wing length: 8.0–9.5 mm

Identification: A small *Eristalis* with a yellow-dusted face that has a slight central, shining black stripe. The most reliable recognition feature is the almost completely yellow (slightly darkened at most) middle tibia, but this must be viewed from the side and not from above. The antennae have very short radiating hairs, a useful character under high magnification but not in the majority of photographs.

Similar species: Confusion with *Eristalis arbustorum* (p. 232) is most likely (see *opposite*).

Observation tips: This species' close resemblance to *E. arbustorum* means that it may be overlooked among its far more numerous congeners unless many individuals are examined. Although mainly a coastal species in Britain, it can occur inland (sometimes in unexpected places). It appears to have a more patchy but substantially more inland distribution in Ireland.

J F M A M J J A S O N D

WING Stigma diffuse and extended

ERISTALINI: *Eristalis*

Separating *E. arbustorum* and *E. abusiva* – It is essential that reliable characters are used rather than those (*e.g.* patterning on the thorax of females) which have proven to be highly unreliable. The most useful feature is the degree to which the middle tibia is darkened. In *E. abusiva* the tibia is pale yellow (at most slightly darkened towards the tip); but in *E. arbustorum* it is strongly darkened towards the apex. Beware, this feature can be misinterpreted if viewed from above. In males, the length along which the eyes are touching is longer in *E. arbustorum* than in *E. abusiva* but this is a somewhat subjective character and is best used in combination with the coloration of the tibiae.

♂ *Eristalis arbustorum* × 6

FACE typically entirely yellow-dusted

HIND LEG metatarsus dark; inflated (see p. 231)

♀ *Eristalis abusiva* × 6

FACE yellow-dusted with a slight central black stripe

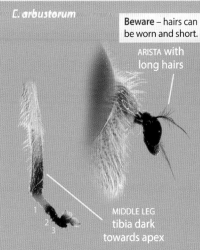

E. arbustorum

Beware – hairs can be worn and short.

ARISTA with long hairs

MIDDLE LEG tibia dark towards apex

E. abusiva

ARISTA with very short hairs

MIDDLE LEG tibia almost completely yellow

Eristalinus

from 3a
p.219

2 British and Irish species (both illustrated)

Dark, shiny, slightly metallic and rather dumpy hoverflies, with distinct dull patches running along the central axis of the abdomen. **The loop in wing vein R_{4+5} and spotted eyes is a combination of characters that is unique among British hoverflies.** The eye spots result from the physical structure of the surface of the eye and are lost in preserved specimens as the eyes dry out. The larvae are of the rat-tailed aquatic type and live among decaying vegetable matter in water.

LC *Eristalinus aeneus*　

LC

Wing length: 6·25–9·5 mm

Identification: The absence of hairs on the lower part of the eye is diagnostic. The abdomen lacks any dull patches and is often slightly bronzy-black in appearance. Males have eyes that are touching. Females (eyes separated) have a relatively shiny thorax.

Similar species: Apart from *Eristalinus sepulchralis*, which is slightly smaller and has hairy eyes, there should be no confusion; males of that species have convergent but well-separated eyes.

Observation tips: An almost entirely coastal species, the larvae of which live in pools with accumulations of rotting seaweed. There are, however, occasional inland records that appear to be increasing in frequency. Although the main flight-period is in midsummer, adults occur for much of the year and both males and females hibernate. Widely distributed around the British and Irish coasts, but most abundant in southern and western locations.

GB: Frequent =
Ir: Local

J F M A M J J A S O N D

LC *Eristalinus sepulchralis*　

LC

Wing length: 6·5–8·0 mm

Identification: The eyes are distinctly spotted and have a full covering of hairs. Both males and females have eyes that are separated. Females have parallel lines of dusting on the thorax. This species tends to be small compared with the *Eristalis* species (*pp. 220–233*) with which it is often seen (but the size is quite variable).

Similar species: Apart from *Eristalinus aeneus*, which is slightly larger and has the lower part of the eye bare, there should be no confusion; in males of that species the eyes touch.

Observation tips: A flower visitor that is found mostly near nutrient-enriched pools and ditches, such as those affected by agricultural run-off. Its larvae live in piles of rotting vegetation in water and also in wet manure. Although widespread across much of Britain and Ireland, this is mainly a lowland species that is commonest in the south of both regions and largely coastal farther north. It seems to have undergone a serious decline over the last decade or so.

GB: Widespread ▽
Ir: Frequent

J F M A M J J A S O N D

ERISTALINI: *Eristalinus*

Separation of British and Irish *Eristalinus* species.

SPECIES	FACE
E. aeneus	UPPER PART OF EYE **hairy**; LOWER PART OF EYE **bare**
E. sepulchralis	UPPER & LOWER PARTS OF EYE **hairy**

EYE obvious junction between lower (bare) and upper (hairy) parts

♂ *Eristalinus aeneus* × **6**

♀ *Eristalinus sepulchralis* × **6**

EYE both lower and upper parts hairy; on this individual, the hairs on the lower part of the eye are dusted with pollen

ERISTALINI: *MALLOTA* | *MYATHROPA* See also *Identifying wasp and bee mimics pp. 62–65*

from 2c
p.218

Mallota

1 British species (illustrated)

One of the genera with a loop in vein R$_{4+5}$ and a very convincing honey-bee mimic. The 'rat-tailed maggot' larvae live in deep, water-filled rot-holes in a variety of trees, large, mature or old trees often providing the most suitable breeding sites.

LC *Mallota cimbiciformis*
LC

Nationally Scarce
GB: Local =
Ir: Not recorded

Wing length: 11·25–12·5 mm

Identification: A large honey-bee mimic with strongly thickened hind femora and a somewhat greenish hue.

Similar species: Most likely to be confused with *Eristalis tenax* (*p. 224*) at first sight, but the enlarged hind femora and greenish hue make separation relatively simple.

Observation tips: Adults are flower visitors which generally stay fairly close to breeding sites. This is a scarce species, with scattered records north to Glasgow and Fife. Adults are probably overlooked and it may be easier to find by searching for larvae in water-filled rot-holes.

J F M A M J J A S O N D

from 3b
p.219

Myathropa

1 British and Irish species (illustrated)

A distinctive yellow-and-black species, often described as a wasp mimic, but not very convincing. One of the genera with a loop in vein R$_{4+5}$; veins R$_1$ and R$_{2+3}$ reach the wing margin separately. It has a distinctively patterned thorax, with the dark area towards the back resembling the 'Batman' symbol. The 'rat-tailed maggot' larvae live in wet hollows containing decaying leaves and twigs.

LC *Myathropa florea*
LC

GB: Widespread ▲
Ir: Widespread

Wing length: 7–12 mm

Identification: The 'Batman' marking on the top of the thorax is distinctive. Well-marked individuals are unmistakable but the patterning is very variable, which can lead to confusion.

Similar species: Poorly marked *Myathropa* can be mistaken for other eristalines. Careful attention should be given to the points at which veins R$_1$ and R$_{2+3}$ enter the wing (separately in *M. florea*, and conjoined and petiolate in *Eristalis* (*p. 220*) and *Eristalinus* (*p. 234*)).

Observation tips: Although breeding sites of this abundant species are most often in woodland, it is a great opportunist and will breed in anything that holds water – such as a plastic container. Adults occur from April to November and visit a wide range of flowers. They also frequently bask on sunny leaves. Widespread across Britain but much scarcer in some parts of Scotland, and with apparently a very patchy distribution in Ireland.

J F M A M J J A S O N D

ERISTALINI: *Mallota* | *Myathropa*

♀ *Mallota cimbiciformis* × **5**

Myathropa identification

distinctive 'Batman' marking on thorax

veins R_1 and R_{2+3} reach the wing margin separately (compare with *Eristalis* (INSET))

Eristalis: 'petiole' beyond the junction of veins R_1 and R_{2+3}

♂ *Myathropa florea* × **5** ♀

ERISTALINI WITH A STRIPED THORAX 1/4

Differentiating features of Eristalini with a striped thorax

GENUS	ANTENNAE	HIND TIBIA
Parhelophilus (below)	Orange	**one** dark ring
Anasimyia (p. 240)		**two** dark rings
Helophilus (pp. 242–245)	Black	**one** dark ring
Lejops (p. 244)		**two** dark rings

Parhelophilus

from 4b
p. 219

3 British and Irish species (2 illustrated)

One of a small group of genera (including *Anasimyia* (p. 240), *Helophilus* (pp. 242–245) and *Lejops* (p. 244) – see table *above*) that have **wings with a loop in vein R4+5** and **pale, longitudinal stripes on the top of the thorax**. Determining the sex is also difficult because in the male the eyes are not touching on the top of the head – the best way to check is to look for a male's genital capsule; the abdomen of females is tapered. Like *Anasimyia*, they have orange antennae, but their hind tibia has only one dark ring whereas *Anasimyia* has two. The two commoner species (*P. frutetorum* and *P. versicolor*) have a general orange appearance that is quite distinctive. Apart from male *P. frutetorum* (in which the tubercle on the underside of the hind femur confirms its identity upon close examination), *Parhelophilus* are difficult to separate, especially females. **Comparison with known specimens is necessary for accurate identification**. The 'rat-tailed maggot' larvae live in decaying plant material at the base of emergent vegetation, especially Bulrush, in ponds and ditches.

Parhelophilus frutetorum/versicolor

LC
LC

GB: Frequent/Frequent ▽
Ir: Local (*P. versicolor*)

Wing length: 7–9 mm

Identification: These two species are very difficult to separate, especially the females. Males can be distinguished in the field using a hand lens – there is a small tubercle on the underside of the hind femur of male *P. frutetorum*; this tubercle is lacking in *P. versicolor*. In general, these species cannot be separated reliably in photos unless they are males and a view of the underside of the hind femora can be obtained.

NOTE: *P. frutetorum* not recorded in Ireland

Similar species: The Nationally Scarce *Parhelophilus consimilis* (N/I) bears some resemblance to these species (it has a distinct dark end to the front tibia) but is more likely to be confused in the field with some *Anasimyia* (p. 240) and *Helophilus* (pp. 242–245) – see table *above*.

Observation tips: Around emergent vegetation in ponds and ditches, especially Bulrush; they are regular flower visitors and can be seen at Yellow Iris. *P. frutetorum* seems to be less faithful to breeding sites and can be found some distance away. The map shows the combined distribution of the two species. *P. versicolor* is the only representative of this species pair in Ireland.

J F M A M J J A S O N D

Parhelophilus species not otherwise covered:

P. consimilis	NS	This species tends to look rather darker and yellower than the other two in the genus and is perhaps more like an *Anasimyia* (p. 240) in the field. It is rare but widely distributed in poor-fen habitats primarily in Wales and as far north as southern Scotland.

ERISTALINI: *Parhelophilus*

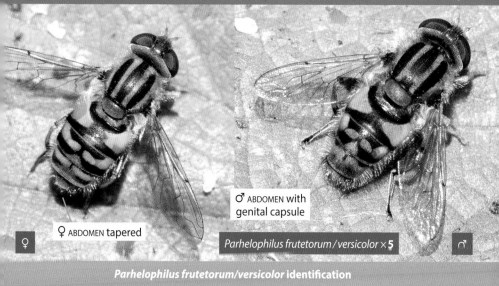

♀ ABDOMEN tapered

♂ ABDOMEN with genital capsule

Parhelophilus frutetorum / versicolor × 5

Parhelophilus frutetorum/versicolor identification

♀ *P. frutetorum* — curved bristles

♀ *P. versicolor* — no curved bristles

♂ *P. frutetorum* HIND FEMUR with curious long-haired tubercle

♂ *P. versicolor* HIND FEMUR no long-haired tubercle

♂ *P. frutetorum* — long-haired tubercle

♂ *P. versicolor* (probably) — no tubercle visible

ERISTALINI WITH A STRIPED THORAX 2/4

from 4b p.219 ## Anasimyia

5 British and Irish species (2 illustrated)

Anasimyia is one of a small group of genera in the tribe Eristalini (including *Helophilus, Lejops* and *Parhelophilus* – see table on *page 238*) that have **wings with a loop in vein R_{4+5} and pale, longitudinal stripes on the top of the thorax**. In both the males and females the eyes are separated. They are most similar to *Parhelophilus* (*p.238*), both having orange antennae, but differ in having two dark rings on the hind tibia, rather than one. *Lejops* (*p.244*) also has two dark rings on the hind tibia but, as *Helophilus* (*pp.242–245*), differ in their dark antennae. *Anasimyia* species are found in wetlands where the aquatic 'rat-tailed maggot' larvae live among rotting material around the base of emergent plants. Adults fly and rest among water plants, visiting nearby flowers and seldom straying far from breeding sites.

Anasimyia lineata

Wing length: 6·25–8·25 mm

Identification: Relatively easy to recognise as it is the only species with the combination of longitudinal stripes on the thorax and a strongly extended face (only *Rhingia* (*p.194*) has a face shaped like this – but it is even more extreme, and the abdomen is orange).

Similar species: Two Nationally Scarce species, *Anasimyia lunulata* (N/I) and *A. interpuncta* (N/I), in both of which the face is not extended.

Observation tips: *A. lineata* can be found at the margins of still or slow-flowing water (including farm and roadside ditches) with emergent vegetation, especially Bulrush. Adults are flower visitors. This is the most widespread and frequent *Anasimyia* in Britain, but appears to have a much more patchy distribution in Ireland.

J F M A M J J A S O N D

Anasimyia contracta

Wing length: 5·0–7·25 mm

Identification: This is perhaps the most likely *Anasimyia* species with hockey-stick-shaped markings to be encountered. It has a somewhat wasp-waisted abdomen, with segment T2 narrower than long.

Similar species: *Anasimyia transfuga* (N/I) has similar abdominal markings, but segment T2 is both parallel-sided and wider than long. Two Nationally Scarce species, *A. lunulata* (N/I) and *A. interpuncta* (N/I), both have abdomen segments T3 and T4 with comma-shaped (not hockey-stick-shaped) markings.

Observation tips: Occurs among emergent vegetation at the margins of still water. Adults visit buttercups and other flowers nearby. Larvae have been found living between the leaf sheaths of rotting plants. Widely distributed in Britain, but rather patchy in Ireland.

J F M A M J J A S O N D

ERISTALINI: *Anasimyia*

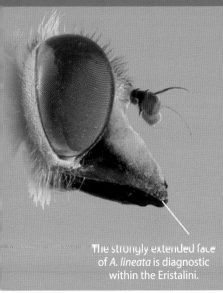

The strongly extended face of *A. lineata* is diagnostic within the Eristalini.

♀ *Anasimyia lineata* × **6**

♂   *Anasimyia contracta* × **6**   ♀

Anasimyia species not otherwise covered:

A. interpuncta	NS	Found largely along the main river valleys of East Anglia and in the Norfolk Broads, Thames marshes and grazing marshes in East Sussex and Somerset. Readily confused with *A. lunulata*, from which it can be separated by facial profile (female) and shape of abdominal markings (difficult and subjective).
A. lunulata	NS	Found mainly in the west, especially west Wales, but also the west coast of Scotland, where it occurs in peatlands. Readily confused with *A. interpuncta* (see *above*).
A. transfuga		Like *A. contracta,* but T2 is not narrowed. This species is much less common than *A. contracta*, but is widely distributed in Wales and the southern half of England. It tends to occur in more shaded situations than other *Anasimyia*.

ERISTALINI WITH A STRIPED THORAX	3/4

Helophilus

from 4a
p. 219

5 British and Irish species (3 illustrated)

Helophilus is one of four genera (with *Anasimyia*, (p. 240) *Lejops* (p. 244) and *Parhelophilus* (p. 238) – see table on *page 238*) that have **wings with a loop in vein R_{4+5} and pale, longitudinal stripes on the top of the thorax**. Like *Lejops*, it has dark antennae, but the hind tibia has only one dark ring (*Lejops* has two). The eyes are separated in both sexes (check the shape of the end of the abdomen – males have a genital capsule) and this is important in identification. The larvae are aquatic 'rat-tailed maggots' found in wet, decaying vegetation in the edges of ponds and ditches.

LC *Helophilus hybridus*
LC

GB: Widespread =
Ir: Widespread

Wing length: 8·5–11·25 mm

Identification: Face yellow with a central black stripe. In males the yellow markings on abdomen segments T2 and T3 are not separated by a black band. Females have black-and-yellow hind tibiae – mostly black, with the yellow portion extending basally (from the femur) along just 25–30% of the length of the tibia; the yellow dusting on the frons extends all the way back to the ocellar triangle.

Similar species: *Helophilus pendulus* (see table *opposite*); *H. trivittatus* (p. 244), in which the face lacks a distinct black stripe and is usually yellow but can have a brownish central stripe.

Observation tips: A wetland species that tends to be found near its breeding habitat, where it visits a variety of flowers. It occurs widely across Britain and Ireland but is less common in the north.

J F M A M J J A S O N D

LC *Helophilus pendulus*
LC

GB: Widespread ▲
Ir: Widespread

Wing length: 8·5–11·25 mm

Identification: Face yellow with a central black stripe. Males have a black band separating the yellow markings on abdomen segments T2 and T3. Females have yellow-and-black hind tibiae – mostly yellow, with the yellow portion extending basally (from the femur) along 50–60% of the length of the tibia; the yellow dusting on their frons stops abruptly well in front of the ocellar triangle.

Similar species: *Helophilus hybridus* (see table *opposite*); *H. trivittatus* (p. 244), in which the face lacks a distinct black stripe and is usually yellow but can have a brownish central stripe.

Observation tips: A common and widespread species which visits a wide range of flowers in a variety of habitats, including gardens, which may be far from breeding sites.

Helophilus **species not otherwise covered:**

J F M A M J J A S O N D

H. affinis	NE	Vagrant: a single record from Fair Isle in 1982. Comparative examination of abdominal and thoracic markings required.
H. groenlandicus	DD	A few Scottish records, mainly from islands of the Inner Hebrides. Comparative examination of abdominal and thoracic markings required.

ERISTALINI: *Helophilus*

Separating *H. hybridus* and *H. pendulus*

SPECIES	*H. hybridus*	*H. pendulus*
ABDOMEN	♂ T2 & T3 SEPARATING BLACK BAND **absent**	♂ T2 & T3 SEPARATING BLACK BAND **present**
HIND TIBIA	♀ black, with **basal 25–30% yellow**	♀ black, with **basal 50–60% yellow**
FRONS	DUSTING **extends to** ocellar triangle	DUSTING **stops short of** ocellar triangle

All British and Irish *Helophilus*, with the exception of *H. trivittatus* (p. 244), have a face with a dark central stripe.

frons dusting
H. pendulus stops abruptly short of ocellar triangle

H. hybridus extends to ocellar triangle

♀ **hind tibia colour**

H. hybridus basal portion 25–30% yellow

H. pendulus basal portion ≥50% yellow

♂ **abdomen**

H. hybridus T2 & T3 yellow 'merges'

H. pendulus ♂ T2 & T3 separated by black band

♂ ABDOMEN genital capsule

♂ *Helophilus hybridus* × 4

Helophilus pendulus × 4

♀ ABDOMEN conical point

♂

ERISTALINI WITH A STRIPED THORAX

Helophilus trivittatus

GB: Widespread
Ir: Frequent

Wing length: 10·25–12·25 mm

Identification: The yellow face, lacking a black central stripe (although there can be a reddish-brown central patch), is distinctive within the genus. The hind tibiae are mainly black. It is also noticeably larger than other *Helophilus* and tends to be a brighter, more lemon-yellow colour, with more extensive yellow markings on the abdomen.

Similar species: Large *Helophilus pendulus* or female *H. hybridus* (both *p. 242*), both of which have a black central stripe on the face.

Observation tips: A migratory species that is encountered most frequently near the coast or along river courses. Numbers fluctuate wildly from year to year. It visits flowers in a wide variety of habitats, including in dry places that are remote from any obvious breeding sites. In Scotland it is predominantly coastal, whilst in Ireland it appears to be mainly western, coastal and absent from much of the east.

J F M A M J J A S O N D

---

from 4a
p. 219

Lejops

1 British species (illustrated)

Lejops is one of a small group of genera (including *Parhelophilus* (p. 238), *Anasimyia* (p. 240) and *Helophilus* (pp. 242–245) – see table on *page 238*) that have **wings with a loop in vein R$_{4+5}$** and **pale, longitudinal stripes on the top of the thorax**. The eyes are separated in both sexes. The antennae are dark and the hind tibia has two dark rings. The larvae are reported to be associated with floating vegetation such as duckweeds until the final instar, when they descend to organic material on the bottom of the water body.

Lejops vittatus

GB: Rare [no trend data]
Ir: Not recorded

Wing length: 8–10 mm

Identification: *Lejops* appears narrower than members of the other genera that have longitudinal stripes on the thorax. Most reminiscent of an *Anasimyia*.

Similar species: *Anasimyia* (p. 240) and *Parhelophilus* (p. 238), which both have orange antennae, and *Helophilus* (pp. 242–245), which has one dark ring on the hind tibia – see table on *page 238*.

Observation tips: A rare species of, predominantly, coastal grazing marshes where it is found in and around beds of Sea Club-rush. It is confined to a few coastal marshes including the Broads (old records), the Thames marshes and the Somerset and Gwent Levels and a few more inland locations. The association with Sea Club-rush may only involve females seeking pollen. In Europe, this species has a wide distribution and is not entirely confined to coastal situations. The larval life-history suggests that not all ditches with Sea Club-rush are suitable and it may be most profitable to seek out ditches with floating mats of Frogbit and duckweeds.

J F M A M J J A S O N D

ERISTALINI: *Helophilus* | *Lejops*

♀ *Helophilus trivittatus* × **4**

face yellow; central stripe, if present, is, at most, reddish-brown

♀ *Lejops vittatus* × **6**

Guide to Merodontini

This tribe comprises three genera, two of which contain a single species, that are very different from each other in appearance. **See the respective species accounts for more information.**

from 4c
p. 56

Eumerus — 5 British and Irish species (all illustrated) [beware further introductions]

Small, dark hoverflies with distinct hair bands on the abdomen and enlarged hind femora. The shape of the re-entrant outer cross-vein is distinctive. The larvae are bulb and rhizome dwellers and at least one species has been introduced to Britain in imported bulbs. Many more species occur in Europe and accidental introductions are always possible. The complex of *E. funeralis*, *E. sogdianus* and *E. strigatus* is most frequently observed but species cannot be identified with confidence from most photographs. Female *E. sogdianus* and *E. strigatus* are inseparable.

LC *Eumerus ornatus*

GB: Local ▽
Ir: Not recorded

Wing length: 4·5–6·25 mm

Identification: Very similar to other *Eumerus* but generally a little larger. The ocelli lie much farther forward on the frons than other similar species but this is a feature that causes considerable confusion. Careful examination is needed to be certain and records are generally not accepted without voucher specimens or photographs that adequately depict the ocelli and ocellar triangle.

Similar species: Confusion with *Eumerus funeralis/sogdianus/ strigatus* (which are generally a little smaller) is most likely: the position of the ocelli is the most reliable means of identification.

Observation tips: This species usually flies close to the ground in dappled light in woodlands. On several occasions it has been found along wet rides with stands of Meadowsweet.

J F M A M J J A S O N D

LC *E. funeralis / sogdianus / strigatus*

GB: Frequent/Rare/Frequent ▽
Ir: Rare/Not recorded/**Local**

Wing length: 3·5–6·25 mm

Identification: Three very similar species that can only be separated by very close examination. *E. funeralis* has a distinct bare area on the underside of the hind femora (hairy in the other two species). It is possible to separate *E. sogdianus* and *E. strigatus* by characters of the male genitalia, but females cannot be separated.

NOTE: *E. sogdianus* not recorded in Ireland

Similar species: *E. ornatus* is very similar but the antennae are usually partially yellow and the ocellar triangle is located farther forward on the frons – a difficult relative character (see *opposite*).

Observation tips: *E. funeralis* is found in urban and rural gardens, visiting a wide range of low-growing flowers. The larvae feed in daffodil and other bulbs where rot is present. *E. strigatus* occurs mainly in open countryside and often visits umbellifers. Its larvae have been found in a variety of bulbs and also iris rhizomes. *E. sogdianus* has been recorded from coastal locations but is poorly known and may be overlooked.

J F M A M J J A S O N D

MERODONTINI: *Eumerus*

♂ *Eumerus ornatus* × 10

E. ornatus

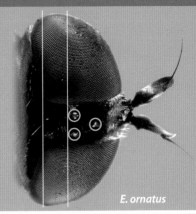

E. funeralis

♀

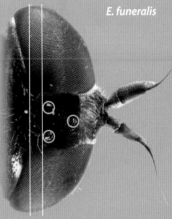

Eumerus funeralis × 10

♂

***Eumerus*: position of ocelli**

The ocelli of *E. ornatus* are positioned farther forward than those of similar *Eumerus* – although this feature requires careful study.

E. funeralis   **hind femora**   *E. strigatus*

bare, polished patch

completely covered in hairs

Eumerus sabulonum

Wing length: 3·25–5·25 mm

Identification: This should be the most straightforward *Eumerus* to identify because the abdomen has reddish markings. A useful pointer, although not conclusive, is that it is predominantly a coastal species.

Similar species: None – no other *Eumerus* in Britain has red markings on the abdomen.

Observation tips: A scarce species which can be found on cliffs and sand dunes along the west coast of England, Wales and into southern Scotland. The larvae have been found to feed in the roots of Sheep's-bit in Denmark, and adults will also visit this plant. This species is often best observed on bare ground at the edge of footpaths or on broken open ground where it seems to be a sun-basker.

Nationally Scarce
GB: Scarce =
Ir: Not recorded

J F M A M J J A S O N D

Merodon

from 2a p.218

1 British and Irish species (illustrated)

A moderately sized polymorphic bumblebee mimic with a loop in vein R_{4+5}. The only other bumblebee mimic with a wing like this is *Eristalis intricaria* (p. 228), but that species has partly pale hind legs (wholly black in *Merodon*). Larvae develop in bulbs, especially daffodils, but it has been recorded from a wide variety of other bulb-forming plants. Many more *Merodon* species occur in Europe, especially around the eastern Mediterranean, and it is quite possible that others could turn up in Britain or Ireland as accidental imports.

Merodon equestris Greater Bulb Fly

GB: Widespread ▲
Ir: Frequent

Wing length: 8·5–10·25 mm

Identification: In addition to the features mentioned *above*, the hind femora are enlarged and have a distinct triangular projection towards the apex. There is a variety of colour forms, each of which mimics a different species of bumblebee (see *page 29*).

Similar species: Other bumblebee mimics such as *Volucella bombylans* (p. 270), *Eristalis intricaria* (p. 228), *Cheilosia chrysocoma* (p. 186), *Criorhina ranunculi* (p. 292), *C. berberina* (p. 294), *C. floccosa* (p. 294) and possibly *Sericomyia superbiens* (p. 266). Photographs of *Eristalis tenax* (p. 224) are not infrequently recorded as *Merodon equestris*, but careful checking of leg colour and the presence or absence of a triangular projection on the hind femora should eliminate confusion.

Observation tips: Believed to have been introduced into Britain in daffodil bulbs imported from Europe towards the end of the 19th century. Adults are frequent in gardens and urban areas across the country, but also occur in the wider countryside. Widespread across Britain but much less abundant in Scotland; it appears to be more coastal in Ireland, perhaps reflecting major urban areas.

J F M A M J J A S O N D

MERODONTINI: *Eumerus* | *Merodon*

♀ *Eumerus sabulonum* × **10**

The reddish base to the abdomen separates *E. sabulonum* from other *Eumerus* species.

♂ *Merodon equestris* × **6**

A variety of colour forms occurs (see *page 29*).

The triangular projection on the enlarged hind femora is unique among the bumblebee mimics.

249

from 2a **Psilota** 1 British species (illustrated)

p. 54 Related to *Merodon* (p. 248) and *Eumerus* (pp. 246–249) but looks very different. It is a steely blue-black hoverfly with bright red eyes, which more closely resembles one of the blue-black muscid flies (e.g. *Hydrotaea*) or the hoverfly *Cheilosia pagana*. **It is the only British hoverfly without any trace of a vena spuria.** Larval biology is unclear but is probably associated with rotting wood in old trees in some way.

LC *Psilota anthracina*
LC

Wing length: 6·0–7·5 mm

Identification: Fairly distinctive – once you realise that it is a hoverfly! The face is flat with a slightly jutting mouth margin, and the wing base and stigma are distinctly yellow.

Similar species: *Cheilosia* species (p. 170) that have indications of silvery patches under certain reflected light, especially *C. impressa* (p. 190), although the presence of a vena spuria and a facial prominence in those species will rule out *Psilota*. Some blue-black muscid and lonchaeid flies may superficially resemble *Psilota* but are all species that are bristly (see inset *below*).

Observation tips: A rare, southern species found most frequently in woodlands with old trees, particularly in the New Forest, Hampshire and Windsor Great Park, Surrey, but also in the Midlands. It is a flower visitor, especially to Hawthorn blossom, with a rather brief main flight-period in May to early June.

Nationally Scarce
GB: Scarce
Ir: Not recorded

J F M A M J J A S O N D

A blue-black muscid fly, *Hydrotaea aenescens*, with which *Psilota* could be confused.

MERODONTINI: *Psilota*

♂ *Psilota anthracina* × **10**

PELECOCERINI: *PELECOCERA*

from
6c # Guide to Pelecocerini

p.58 Formerly two genera, *Pelecocera* and *Chamaesyrphus* are now treated as a single genus, *Pelecocera*.
Easily recognised by the very distinctive large antennae and thickened arista (see *opposite*).

Pelecocera 3 British species (2 illustrated)

Small yellow-and-black hoverflies with unusual antennae that are rather large in proportion to the face. The eyes are separated in both sexes. Adults visit low-growing flowers in ericaceous habitats in coniferous woodland in the Scottish Highlands and on heathland in southern England. The larvae have been found in the fruiting bodies of fungi such as *Rhizopogon luteolus*.

LC *Pelecocera scaevoides*   Nationally Scarce

GB: Rare ▽

Wing length: 4·0–6·25 mm

Ir: Not recorded

Identification: A small species with yellow-and-black abdominal markings and a large 3rd antennal segment with an upright thickened arista (see inset, *opposite*). The yellow on the abdomen has no grey dusting, and there is a bristle immediately in front of the wing base.

Similar species: *Pelecocera caledonicus* (N/I), which lacks a bristle in front of the wing base. *P. scaevoides* may be overlooked as a *Melanostoma* (*p. 94*) or possibly a *Platycheirus* (*pp. 75–93*) in the field but those genera have much smaller antennae. *P. tricincta* should not cause confusion as it is restricted to southern England.

Observation tips: Both *P. scaevoides* and *P. caledonicus* are confined to Caledonian pine forest and conifer plantations in central and northern Scotland and can be found visiting Tormentil and other yellow flowers along paths and rides.

|||| |
J F M A M J J A S O N D

LC *Pelecocera tricincta* Nationally Scarce

GB: Scarce ▲

Wing length: 3·5–5·25 mm

Ir: Not recorded

Identification: A small species with yellow-and-black abdominal markings and a large half-moon-shaped 3rd antennal segment; the thickened arista forms an extension beyond the antennal segment (see inset, *opposite*). Within its range, readily recognised by the unusual shape of the antennae.

Similar species: Can be overlooked as *Melanostoma* (*p. 94*) or possibly a *Platycheirus* (*pp. 75–93*) in the field but those genera have much smaller antennae. *P. scaevoides* should not cause confusion as it is restricted to northern Scotland

Observation tips: A southern heathland species which is normally found visiting yellow flowers such as Tormentil and hawkweeds in the grassy margins of tracks and paths. Found on the heaths of Dorset, Hampshire and Surrey, and less commonly in Devon and West Sussex. It can be reasonably common within its limited range.

J F M A M J J A S O N D

PELECOCERINI: *Pelecocera*

Pelecocera scaevoides × 10 ♂

Pelecocera antennae

P. scaevoides

ARISTA short and somewhat vertical

P. tricincta

ANTENNAE half-moon-shaped; ARISTA a thickened section beyond segment 3

♂ *Pelecocera tricincta* × 10

Pelecocera species not otherwise covered:

| P. caledonicus | VU | Difficult to separate from *P. scaevoides*. Culbin Sands on the Moray Firth is its main locality (most recently in 2006). There have been records from four other Caledonian pine forest sites in northern Scotland. |

Guide to Pipizini

from 6a p.58

The five genera within this tribe are predominantly black, although some bear round yellow markings on abdomen segment T2. Many species are small to medium-sized and rest in a distinctive manner on leaves and flowers with their wings held out in a rather delta-like shape. The humeri are hairy and in most cases this is readily apparent. **The most useful character for separating this tribe from all others is the flat face, covered in long, drooping hairs, with no evidence of a central prominence and without a distinctly projecting mouth margin.** It is a tribe that is frequently overlooked by recorders, not least because identification of the species can be very challenging. Pipizines are rarely identifiable from photographs with any certainty and all such identifications should be treated with caution!

Pipizini face | The flat face covered in drooping hairs, with no central prominence and without a distinctly projecting mouth margin typifies members of the Pipizini.

Pipiza austriaca

Psilota anthracina

Beware – *Psilota anthracina* [**Merodontini**] (*p. 250*) may also key out here. It is a shining bluish-black hoverfly that has a flat face with a strongly pointed mouth-edge. However, *Psilota* lacks a vena spuria (see *page 54*) so differentiation from a pipizine hoverfly should be straightforward. *Psilota* is easily confused with some blue-black muscid flies and so great care should be taken with its identification.

PIPIZINI

1

a) **Abdomen with only two obviously visible segments (T4 is not apparent)**

▶ *Triglyphus primus* p. 264

b) **Abdomen with three or four segments visible**
▶ **2**

2 from 1b

a) **Upper-outer cross-vein of wing curved and more upright**

Trichopsomyia females have yellow spots on abdomen segment T2; otherwise these are small and entirely black hoverflies.

Legs with pale hairs
▶ *Pipizella* p. 262
Hind tibia with black hairs
▶ *Trichopsomyia* p. 264

b) **Upper-outer cross-vein of wing strongly sloping and less curved**

In general, *Pipiza* are larger (*Heringia* tend to have a wing length less than 6 mm), but small *Pipiza* do occur. Male *Neocnemodon* have a distinctive spur on their hind trochanter (see *page 261*). *Pipiza* often have yellow spots on the abdomen.

▶ *Pipiza* pp. 256–259
▶ *Heringia* and *Neocnemodon* p. 260

See accounts for more information on identification of these difficult genera.

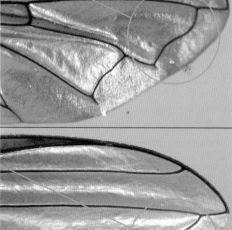

PIPIZINI: *PIPIZA*

from
2b
p.255

Pipiza

7 British and Irish species (3 illustrated)

Medium-sized black hoverflies with a flat, hairy face and often with a wing cloud. Many of the species have yellow spots on abdomen segment T2. **However, they are very variable and, in many cases, it is difficult to be absolutely certain of their identity.** Indeed, of all the genera within the family Syrphidae, this is regarded as one of the most intractable. For this reason, only the more readily identifiable species are dealt with here. The larvae feed on aphids, often in enclosed situations such as aphid-induced galls and leaf curls. A major revision of the genus *Pipiza* in 2013, including DNA analysis, resulted in some changes to species names (*P. bimaculata* is now *P. notata*, and *P. fenestrata* is now *P. fasciata*), but the overall composition of the British and Irish fauna remains the same.

Identification of *Pipiza* is based upon a range of characters – most critically, the distribution of different coloured hairs on the face, abdomen segments T4 and T5, and margins of the abdomen segments. The degree of darkening of the wings is also used, but requires comparison with voucher specimens as it is very easy to misinterpret. DNA analysis has also been used and although this has resolved a few issues, it has also raised others.

LC *Pipiza austriaca*
LC

GB: Frequent =
Ir: Local

Wing length: 6–8 mm

Identification: One of the largest species of the genus and fairly easy to recognise because it has ridges on the underside of the hind femora, towards the apex, which gives them a thickened appearance. The abdomen is black in both sexes (any apparent markings are due to bands of hair, not coloured spots), and the wings have strong but diffuse black clouds.

Similar species: Most likely to be confused with *Pipiza noctiluca* (*p. 258*) and *P. fasciata* (N/I) and requires comparative evaluation. (See also comparison table *opposite* for similar genera.)

Observation tips: Usually found visiting flowers such as buttercups and Hogweed in meadows, hedgerows and woodland rides and edges. Widespread in England and Wales, and north to the Scottish lowlands, but not usually abundant and seemingly declining in south-east England. There appears to be a corresponding southern bias in its Irish distribution.

J F M A M J J A S O N D

Pipiza species not otherwise covered:

P. notata	Widely scattered and most frequent in southern England; just a single pre-2000 Irish record. Very difficult to distinguish from *P. noctiluca* (*p.258*).
P. fasciata	A few, mostly old, records mainly from south of the Humber, but with a scattering north to Fife. Difficult to distinguish from *P. noctiluca* (*p.258*) but often larger.
P. festiva	A single Irish record, the origin of which is seemingly uncertain.
P. lugubris NS	Widely distributed across England and Wales north to the Humber, but mostly in southern central England. Sometimes recognisable from photographs due to the highly defined wing cloud (a subjective character that requires familiarity with specimens).

Pipiza notata, P. fasciata, P. festiva, P. lugubris and *P. noctiluca* (*p.258*) are part of an extremely difficult complex of species and are avoided by many recorders. Consequently, their distribution and status are not very well established.

PIPIZINI: *Pipiza*

Differentiating features of the Pipizini

GENUS	ABDOMEN	WINGS		OTHER FEATURES
Triglyphus (p. 264)	2 visible segments			
Pipiza (pp. 256–259)		UPPER-OUTER CROSS-VEIN **strongly sloping and less curved**		HEAD ♀ **triangular dust bars**; ♂ frons less swollen (see *page 261*)
Heringia (p. 260)				HEAD ♀ **narrow dust bars**; ♂ frons more swollen (see *page 261*)
Neocnemodon (p. 260)	3–4 visible segments			HEAD ♀ **dust bars tiny or absent**; ♂ projecting spur on hind trochanter
Pipizella (p. 262)		UPPER-OUTER CROSS-VEIN **curved and more upright**		LEGS **pale hairs** ABDOMEN ♀ wholly black
Trichopsomyia (p. 264)				LEGS **black hair on hind tibia** ABDOMEN ♀ yellow spots on T2

The flat, hairy face of *Pipiza* – a feature shared by all the pipizines (see also *page 261*).

The underside of the hind femur has a thickened ridge – a feature that makes *P. austriaca* relatively easy to identify.

♂ *Pipiza austriaca* × 10

Pipiza luteitarsis

GB: Frequent ▽
Ir: Rare

Wing length: 6·5–8·0 mm

Identification: Most *Pipiza* have some degree of yellow on the tarsi, but *P. luteitarsis* has extensively yellow tarsi on the front and middle legs (although in some individuals the outermost tarsal segment is darkened). Both sexes have yellow spots on abdomen segment T2, but they are smaller in males and can be rather vague.

Similar species: Most likely to be confused with *Pipiza noctiluca* and *P. fasciata* (N/I) (but see comparison table on *page 257* for similar genera).

Observation tips: Most frequently found around elm trees, where the larvae feed upon aphids, which causes the leaves to curl. The adults bask on sunlit leaves and males hover close to the foliage, defending sun spots. An uncommon, predominantly southern species, with few Scottish records. It is known from just 16 records in Ireland.

J F M A M J J A S O N D

Pipiza noctiluca

GB: Frequent ▽
Ir: Local

Wing length: 6·5–8·0 mm

Identification: An enormously variable species which may prove to be a complex of several species. The size, intensity of wing shading and tarsal coloration (from largely dark to having yellow tarsi on the middle leg) all vary between individuals. The female abdomen typically has a pair of yellow spots, whilst the male is usually unmarked, but the extent of spotting also varies greatly between individuals and several different forms have been described (a male with yellow spots on T2 is illustrated).

Similar species: Its size range means that it could be confused with *Pipiza notata*, which is often very small and is almost identical; separation depends upon comparative material and often causes considerable difficulty. Confusion with *P. austriaca* (*p. 256*), *P. fasciata* (N/I) and *P. luteitarsis* also require comparative evaluation. The genera *Pipizella* (*p. 262*), *Neocnemodon* (*p. 260*), *Heringia* (*p. 260*), *Trichopsomyia* (*p. 264*) and *Triglyphus* (*p. 264*) are most likely to be mistinetrpreted as small examples of *Pipiza noctiluca* (see comparison table on *page 257*).

J F M A M J J A S O N D

Observation tips: A regular flower visitor, especially to umbellifers, but equally likely to be found basking on sunlit leaves. Widespread and fairly common throughout most of southern Britain, as far north as about the Lake District, but scarcer farther north. In Ireland, it appears to be predominantly a southern species.

PIPIZINI: *Pipiza*

♂ *Pipiza luteitarsis* × 10

front tarsi mainly yellow

♂ *Pipiza noctiluca* × 10

from 2b
p. 255

Heringia
2 British and Irish species (1 illustrated)

Small black hoverflies that have a flattened, hairy face. The British list comprises two species, *H. heringi* and *H. senilis*, but there is growing scepticism about the status of the latter, which may prove to be only a form of *H. heringi*. The larvae feed on aphids within leaf galls.

LC *Heringia heringi*
LC
GB: Local =
Ir: Rare

Wing length: 5·5–6·25 mm

Identification: Males lack a spur on the hind trochanter (present in the closely related genus *Neocnemodon*). Females have narrow dust bars on the face (absent in *Neocnemodon*).

Similar species: *Heringia senilis* (N/I), *Neocnemodon* (below), *Pipizella* (p. 262), *Trichopsomyia* (p. 264), *Triglyphus* (p. 264) and some small *Pipiza* (p. 256) in which the males have a less swollen frons and no spurs or leg modifications and females have triangular dust bars (see comparison table on *page 257*).

Observation tips: Adults are most likely to be found basking on sunlit leaves and do not usually visit flowers. The larvae feed on aphids within galls on poplars, elms and willows. Collecting recently fallen spiral galls on the leaf petiole of Lombardy and native Black Poplars to secure larvae is a way of finding this species. There are just five records from Ireland.

J F M A M J J A S O N D

from 2b
p. 255

Neocnemodon
5 British and Irish species (1 illustrated)

Small black hoverflies that have a flattened, hairy face and modifications on the legs of some males. **All *Neocnemodon*, as with the majority of the Pipizini, require microscopic examination to confirm identification.** The larvae feed on aphids in leaf curls.

LC *Neocnemodon pubescens*[1] / *vitripennis*
LC
Nationally Scarce [1]
GB: Scarce/Local =
Ir: Not recorded/Rare

Wing length: 4·5–6·0 mm

Identification: Males have a spur on the hind trochanter (absent in *Heringia*). Females lack dust bars on the face (present in *Heringia*). Males are recognised by modifications of the front legs and genitalia characters. **Females are problematic, identification relying upon subjective interpretations of hair patterns on the frons, making separation of the species less reliable.**

NOTE: *N. pubescens* not recorded in Ireland

Similar species: Other *Neocnemodon*, *Heringia* (above), *Pipizella* (p. 262), *Trichopsomyia* (p. 264), *Triglyphus* (p. 264) and some small *Pipiza* (p. 256) (see table on *page 257*).

Observation tips: *N. vitripennis* is the most frequently encountered *Neocnemodon* in Britain and has been captured in numbers in water traps placed in bramble patches; this may indicate its preferred habitat. Adults do not usually visit flowers. *N. pubescens* is associated with conifer plantations. Both species are most frequent in southern England (the map shows their combined distribution). *N. vitripennis* is the only representative of this species pair in Ireland but to date there have been just 13 records.

J F M A M J J A S O N D

PIPIZINI: *Heringia* | *Neocnemodon*

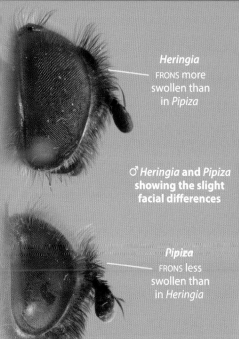

Heringia FRONS more swollen than in *Pipiza*

♂ *Heringia* and *Pipiza* showing the slight facial differences

Pipiza FRONS less swollen than in *Heringia*

♂ *Heringia heringi* × 10

UPPER-OUTER CROSS-VEIN strongly sloping and only slightly curved in *Pipiza*, *Heringia* and *Neocnemodon*

The spurs on the hind trochanters that typify *Neocnemodon*

♀ *Neocnemodon* sp. × 10

Heringia and Neocnemodon species not otherwise covered:

Heringia senilis		Very similar to *H. heringi* (separated on details of the male genitalia) and not regarded as a distinct species by many European authors. Very few, scattered British records.
Neocnemodon brevidens	NS	These are rare and little-known species, although the lack of records may reflect many recorders avoiding them because they are so difficult to identify.
N. latitarsis	NS	
N. verrucula	DD	

from 2a *Pipizella*
p. 255

3 British and Irish species (2 illustrated)

Small black hoverflies with rather elongate antennae and distinctive wing venation due to the shape of the outer cross-vein (see *page 255*); they also have pale hairs on the anterior face of the tibiae. **Identification in the field or from photographs is not straightforward – best identified by microscopic examination of the male genitalia.** This is one of several genera that mimic small solitary bees, especially in their behaviour. The larvae feed on root-feeding aphids.

Similar species: The three *Pipizella* species look very similar to each other (see *above*). *Heringia* (p. 260), *Neocnemodon* (p. 260), *Pipiza* (p. 256), *Trichopsomyia* (p. 264) and *Triglyphus* (p. 264) species may also cause confusion (see comparison table on *page 257*).

LC *Pipizella viduata*

GB: Widespread =
Ir: Local

Wing length: 3·75–6·25 mm

Identification: The shape of the male genital capsule is the most reliable means of confirming identification (readily teased out and twisted round to see the features). The length of hairs on the hind tibiae are useful in both males and females (see *opposite*). The 3rd antennal segment is shorter than that of the other two species and the arista is darker (both are subjective features and difficult to photograph).

Observation tips: Favours dry grassland and woodland rides, where adults bask on leaves or patches of bare ground and visit flowers such as bedstraws and umbellifers. *P. viduata* is widespread across Britain, but is probably often overlooked. It is the only *Pipizella* species recorded in Ireland to date.

J F M A M J J A S O N D

LC *Pipizella virens*

GB: Frequent =
Ir: Not recorded

Wing length: 3·75–6·25 mm

Identification: The shape of the male genital capsule is crucial for separating this species from both *Pipizella viduata* and *P. maculipennis*. The 3rd antennal segment is longer in *P. virens* than in *P. viduata* (see *opposite*) but this is a very subjective and unreliable character. *P. virens* is part of a complex, including *P. maculipennis*, in which only the males can be identified.

Observation tips: Occurs mainly to the south of a line between the Humber and the Mersey, and absent from Ireland. It is a leaf-basker that visits open white flowers, especially umbellifers.

J F M A M J J A S O N D

Pipizella species not otherwise covered:

P. maculipennis	NS	Rare species with about 20 records since 1980 scattered across England south of Warwickshire (not recorded in Ireland).

PIPIZINI: *Pipizella*

UPPER-OUTER CROSS-VEIN curved and more upright in both *Pipizella* and *Trichopsomyia*

♀ *Pipizella viduata* × **10**

Separation of *P. viduata* and *P. virens*

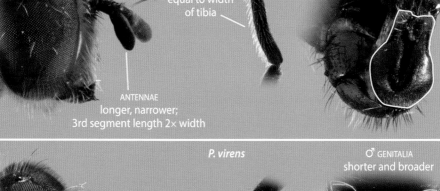

P. viduata

HIND TIBIA hairs on outside face shorter than or equal to width of tibia

♂ GENITALIA longer and narrower

ANTENNAE longer, narrower; 3rd segment length 2× width

P. virens

HIND TIBIA hairs on outside face longer than width of tibia

♂ GENITALIA shorter and broader

ANTENNAE longer, narrower; 3rd segment length 2½× width

Trichopsomyia

from 2a
p. 255

2 British and Irish species (1 illustrated)

Small black hoverflies resembling *Pipizella* (*p. 262*), but the females have round yellow spots on abdomen segment T2, and both sexes have black hair on the hind tibia (*Pipizella* is entirely black and has pale hairs on the anterior face of the tibiae).

LC *Trichopsomyia flavitarsis*

LC

Wing length: 4–6 mm

Identification: The wing venation and general shape are similar to *Pipizella*, but the female T2 yellow markings are more like *Pipiza*. The long, black hairs on the hind tibia are distinctive, but can only be seen by careful examination under high magnification.

Similar species: *Trichopsomyia lucida* (N/I); *Pipiza*, particularly *P. noctiluca/notata* (pp. 258/256), *Pipizella* (p. 262), *Neocnemodon* (p. 260) and *Heringia* (p. 260) species (see comparison table on *page 257*).

Observation tips: Mainly a western and upland species in Britain, found in damp, rushy pastures, especially where Jointed Rush is abundant. It occurs also in some lowland situations, such as acid heaths and rich fens, and, surprisingly, on some wet calcareous grasslands. In Ireland, it seems to be most frequent on the west coast but absent or rare in both the north-east and the south-east.

GB: Frequent ▽
Ir: Frequent

J F M A M J J A S O N D

Triglyphus

from 1
p. 255

1 British species (illustrated)

A tiny black hoverfly resembling *Pipizella* (*p. 262*) which, on close examination, is **immediately identified by the structure of its abdomen, having only two segments visible** (see *page 255*). All other hoverflies have at least three or four abdomen segments visible. The larvae feed in aphid galls on Mugwort.

LC *Triglyphus primus*

LC

Wing length: 4.25–5.0 mm

Identification: Often occurs among other species from the tribe Pipizini, especially *Pipizella*, and is easily overlooked. The unusual arrangement of the abdomen segments should make it readily recognisable, but this feature does require careful examination under magnification (although it can sometimes be discerned in good photographs).

Similar species: *Heringia* (p. 260), *Neocnemodon* (p. 260), *Pipizella* (p. 262) and *Pipiza* (p. 256) species (see comparison table on *page 257*).

Observation tips: Scarce, with a southern distribution. Most records are from localities where Mugwort is abundant (*e.g.* rough grassland, urban waste ground and old industrial sites). Adults tend to visit umbellifers, particularly Wild Parsnip and Wild Carrot.

Nationally Scarce
GB: Local =
Ir: Not recorded

J F M A M J J A S O N D

PIPIZINI: *Trichopsomyia* | *Triglyphus*

♀ *Trichopsomyia flavitarsis* × 10

♀ showing round spots

UPPER-OUTER CROSS-VEIN
curved and more upright in both *Pipizella* and *Trichopsomyia*

Trichopsomyia species not otherwise covered:

T. lucida	Discovered in Britain in 2006 in a cemetery in London. It may be confused with yellow-spotted *Pipiza* (p. 256) and could be overlooked.

♀ *Triglyphus primus* × 10

SERICOMYIINI: *SERICOMYIA* See also *Identifying wasp and bee mimics pp. 62–65*

from
5b Guide to Sericomyiini

p.57 The single genus within this tribe comprises large, colourful hoverflies with **plumose aristae and the outer cross-veins of the wings not re-entrant**. *Volucella* (pp. 270–275) have similar aristae, but the re-entrant outer cross-veins distinguish that genus.

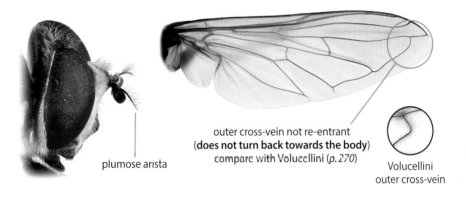

plumose arista

outer cross-vein not re-entrant
(does not turn back towards the body)
compare with Volucellini (p. 270)

Volucellini
outer cross-vein

Sericomyia   3 British and Irish species (all illustrated)

Large bumblebee and wasp mimics with plumose aristae. Confusion is most likely with the genus *Volucella* (pp. 270–275) but *Sericomyia* lacks the re-entrant outer cross-vein of that genus. They can be readily identified from photographs. The larvae are aquatic and live in peaty pools and ditches.

LC *Sericomyia superbiens*
LC
Wing length: 10·0–13·5 mm

GB: Frequent ▽
Ir: Frequent

Identification: A large, buff-coloured, bumblebee mimic with distinct dark wing markings and strongly plumose aristae.

Similar species: The buff form of *Volucella bombylans* (p. 270) is the most likely source of confusion but the wing venation (outer cross-vein re-entrant) is distinctive and different. It may also be confused with the pale form of *Criorhina berberina* and perhaps even *C. floccosa* (both p. 294), but these have bare aristae and do not have distinct dark wing markings. Confusion with some forms of *Merodon equestris* (p. 248 and p. 29) has been noted from photographs, but that species has a different wing venation (loop in vein R4+5) and a triangular projection on the hind femora.

Observation tips: Damp meadows and clearings in the north and west, although it also occurs in Norfolk. Its flight-period is unusually late, peaking in late summer to early autumn, when it tends to visit blue and purple flowers such as knapweeds and Devil's-bit Scabious. In Ireland, it has a somewhat patchy range, with modelled distribution suggesting that it is most abundant along the border between the Republic of Ireland and Northern Ireland.

J F M A M J J A S O N D

NOTE: Formerly known as *Arctophila superbiens*.

SERICOMYIINI: *Sericomyia*

Sericomyia superbiens × 5

Sericomyia lappona

Wing length: 10.0–13.5 mm

Identification: The narrow, whitish markings on the abdomen and the reddish scutellum combine to make this species easily identifiable.

Similar species: Has been misidentified as *Eristalis cryptarum* (*p. 228*) but that species' very different wing venation (loop in vein R_{4+5}) should eliminate any confusion.

Observation tips: Occurs mainly in May and June, earlier than *S. silentis*, but the flight-periods do overlap. Unlike *S. silentis*, *S. lappona* does not normally move far from breeding locations and is usually seen visiting flowers on or near bogs. In Britain, it is primarily a northern and western species but also occurs on southern heathlands. In Ireland, it seems to predominate on the west coast although it is generally widely distributed.

J F M A M J J A S O N D

Sericomyia silentis

Wing length: 9.5–14.0 mm

Identification: A very large, black-and-yellow wasp mimic that is unlikely to be confused with any other species.

Similar species: Has been misidentified as *Volucella inanis* and *V. zonaria* (both *p. 274*) – the large, black-and-yellow marked members of that genus – but the wing venation of *Volucella* (outer cross-vein re-entrant) is distinctive and different (see *page 266*).

Observation tips: This hoverfly appears to be very mobile and can be found well away from obvious breeding sites. Adults seem to prefer red and purple flowers, such as thistles and knapweeds, but a wide range of flowers may be visited. It is a widespread and abundant upland, northern and western species that favours acid wetlands. In lowland and eastern regions of Britain it occurs mainly on heathlands and other acidic habitats. It is one of the most frequently recorded hoverflies in Ireland.

J F M A M J J A S O N D

Note: There are regular submissions to iRecord of this species under the name *Eristalis cryptarum* (see *page 228*). This problem arises because the colloquial name Bog Hoverfly has been applied to both species in various books but only relates to *E. cryptarum* in the iRecord species dictionary (hence use of the name Bog Hoverfly is not encouraged here).

SERICOMYIINI: *Sericomyia*

♀ *Sericomyia lappona* ×5

♂ *Sericomyia silentis* ×5

VOLUCELLINI: *VOLUCELLA*   See also *Identifying wasp and bee mimics pp. 62–65*

from 4 Guide to Volucellini

p. 56 This tribe of large hoverflies is represented in Britain and Ireland by a single genus, *Volucella*, which can be easily recognised by the **plumose aristae and wings with the outer cross-vein re-entrant**. Members of the Sericomyiini (*pp. 266–269*) have similar aristae, but the outer cross-vein is not re-entrant.

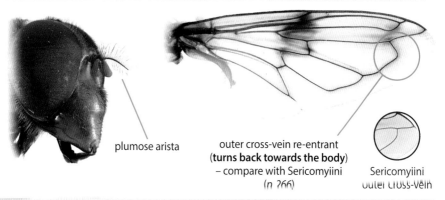

plumose arista

outer cross-vein re-entrant
(turns back towards the body)
– compare with Sericomyiini
(p. 266)

Sericomyiini
outer cross-vein

Volucella   5 British and Irish species (all illustrated)

Among the largest British hoverflies and instantly recognisable. The five British species (only two of which have been recorded in Ireland) are quite diverse in general appearance and are fairly easy to distinguish. The larvae of all except *V. inflata* (*p. 272*) live in the nests of social wasps or bumblebees.

LC *Volucella bombylans*

| GB: Widespread ▲ |
| Ir: Widespread |

Wing length: 8–14 mm

Identification: A polymorphic bumblebee mimic. The plumose aristae and the distinctive *Volucella* wing venation (outer cross-vein re-entrant) (see *above*) separate it from other bumblebee mimics. Two forms are common: a black form with a red-orange 'tail' resembling the bumblebee *Bombus lapidarius*, and a black-and-yellow form which resembles *B. lucorum* (see insets *opposite*). A third, entirely buff-coloured form, resembling *B. pascuorum* (see *page 294*), is much scarcer.

Similar species: The most likely confusion species is *Sericomyia superbiens* (*p. 266*), possibly also *Criorhina berberina* (*p. 294*), *C. floccosa* (*p. 294*), *C. ranunculi* (*p. 292*), *Merodon equestris* (*p. 248*) and *Pocota personata* (*p. 296*), but all lack the distinctive *Volucella* wing venation.

Observation tips: A widespread and common visitor to a wide range of flowers. Males are highly territorial, defending a sunlit leaf and making forays to investigate possible mates. The larvae live in bumblebee nests, where they are scavengers among debris in the bottom of the nest. Although this species occurs throughout Britain, it seems to be absent from some northerly parts of Ireland.

J F M A M J J A S O N D

Volucella inflata

GB: Frequent ═
Ir: Not recorded

Wing length: 11·0–12·75 mm

Identification: A medium-sized, almost globular hoverfly with golden-orange markings on abdomen segment T2 and strong black wing clouds.

Similar species: *Volucella pellucens*, which has similar-shaped, but white, abdominal markings. The markings of some male *V. pellucens* can be strongly yellowish (see photo *below*), occasionally leading to confusion, but they are never anything like the golden colour of *V. inflata*.

Observation tips: A woodland species which is locally abundant in southern England and south Wales. The adults are regular flower visitors, favouring Bramble and shrubs such as Dogwood and Wild Privet. Males fly rapidly around and through such bushes seeking mates. Unlike other British *Volucella*, the larvae inhabit sap runs.

J F M A M J J A S O N D

Volucella pellucens Great Pied Hoverfly

GB: Widespread ═
Ir: Widespread

Wing length: 10·0–15·5 mm

Identification: One of the most obvious hoverflies in Britain and Ireland, with extensive white markings on abdomen segment T2 contrasting with the otherwise black thorax and abdomen.

Similar species: Dark males of *Volucella pellucens* are sometimes misidentified as *V. inflata*, which is slightly more globular in appearance and has orange abdominal markings. *Leucozona lucorum* (p. 116), *Cheilosia illustrata* (p. 176) and *Eristalis intricaria* (p. 228) have vaguely similar colour patterns but lack the distinctive *Volucella* wing venation (outer cross-vein re-entrant). *L. lucorum* is especially frequently misidentified as *V. pellucens* and early spring records should therefore be treated with caution

Observation tips: Widespread and abundant. Adults are generally found in sheltered situations such as woodland rides and tree-lined paths. Males hover at around head height and defend a beam of sunshine. They dart off to investigate intruders or possible females and then return to hover in the same spot. Both sexes visit a wide range of flowers and are often seen in gardens. Larvae live in the nests of a range of social wasps where they are scavengers among the debris in the bottom of the nest cavity. Likely to be found in all parts of Britain except the north of Scotland and the outer islands. It seems to be less well recorded in northerly parts of Ireland.

J F M A M J J A S O N D

Some male *Volucella pellucens* have very dark markings on abdomen segment T2 and can be mistaken for *V. inflata*.

VOLUCELLINI: *Volucella*

♀ *Volucella inflata* × **4**

♂ *Volucella pellucens* × **4** ♀

VOLUCELLINI: *VOLUCELLA* See also *Identifying wasp and bee mimics pp. 62–65*

Volucella inanis

GB: Widespread
Ir: Not recorded

Wing length: 12·25–14·25 mm

Identification: A large yellow-and-black wasp mimic that is only likely to be confused with *Volucella zonaria* (see comparison *opposite* and *below*).

Similar species: *Volucella zonaria*, and possibly *Sericomyia silentis* (*p. 268*), although the difference in wing venation should be easy to see.

Observation tips: Widely distributed across southern England and north to Yorkshire. It has undergone a dramatic expansion in range since about 1995, having previously been confined to southern England, especially the London area. Larvae live in the nests of ground-nesting social wasps, where they feed on the wasp grubs. Since wasps often build their nests in buildings, this hoverfly frequently turns up indoors, often being seen on windows. Adults visit a wide range of flowers in mid- to late summer.

J F M A M J J A S O N D

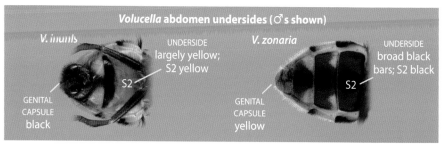

Volucella abdomen undersides (♂'s shown)

V. inanis — UNDERSIDE largely yellow; S2 yellow — S2 — GENITAL CAPSULE black

V. zonaria — UNDERSIDE broad black bars; S2 black — S2 — GENITAL CAPSULE yellow

Volucella zonaria Hornet Hoverfly

GB: Widespread
Ir: Not recorded

Wing length: 15·5–19·5 mm

Identification: The largest British hoverfly and often described as a Hornet mimic. It should not be mistaken for any hoverfly other than *Volucella inanis* (see comparison *opposite* and *above*).

Similar species: *Volucella inanis*. Reports of *V. zonaria* from northern England and southern Scotland have proved to be *Sericomyia silentis* (*p. 268*).

Observation tips: Widespread and increasingly abundant in southern England south of a line between Lancashire and Humberside. Usually seen visiting garden flowers such as *Buddleja*. A relatively recent colonist that arrived on the south coast of England in the late 1930s. It became established, mainly in the London area, resulting in a steady stream of notes in the entomological press from the 1940s through to the 1970s. Consequently, its distribution changes are among the best documented of any hoverfly. Since about 1995, it has expanded its range rapidly, and this trend is continuing, with occasional records from the central lowlands of Scotland. Larvae live in the nests of social wasps that build in tree cavities, including the Hornet. They are scavengers among the debris in the bottom of the nest cavity.

J F M A M J J A S O N D

VOLUCELLINI: *Volucella*

♀ *Volucella inanis* ×4

SCUTELLUM
dull yellowish

ABDOMEN SEGMENT T2
with yellow markings

♀ *Volucella zonaria* ×4

SCUTELLUM
chestnut

ABDOMEN SEGMENT T2
with chestnut markings

Guide to Xylotini

from 3d p.55

A heterogeneous tribe composed of ten genera, seven of which contain just a single species. The humeri are hairy, a feature that can be seen easily due to the head and the front of the thorax being well separated. Many of the genera are instantly recognisable – with practice. Some have distinctly enlarged and ornamented hind femora; others are bumblebee mimics; and yet others resemble ichneumon wasps or sawflies of the genus *Macrophya*. **The relative position of the vein R–M in relation to the discal cell is diagnostic: it joins the upper margin of the discal cell at a point at or beyond the middle of this cell** (see *page 278*).

1

a) Distinctive species:

Metallic greenish-black hoverfly with orange-brown wings and orange legs	Abdomen black-and-orange
▶ *Caliprobola speciosa* p. 280	▶ *Blera fallax* p. 278

Both *Blera* and *Caliprobola* are rare.

b) Narrow-bodied species with grey-dusted thorax side:

Hind femora greatly enlarged, without flange	Hind femora greatly enlarged, with flange
▶ *Syritta pipiens* p. 290	▶ *Tropidia scita* p. 290

c) Bumblebee mimics ▶ **2**

d) Honey-bee mimics ▶ **3**

e) Sawfly mimics ▶ **4**

XYLOTINI

BUMBLEBEE MIMICS

2 from 1c

a) **Distinctive face with elongated, downward-pointing mouth edge**
   ▶ *Criorhina* pp. 292–295

b) **Head small in comparison to body**
   ▶ *Pocota personata* p. 296

HONEY-BEE MIMICS

3 from 1d

a) **Hind femora not enlarged: Distinctive face-shape**
   ▶ *Criorhina asilica* p. 292

b) **Hind femora enlarged:**

   Underside of thorax bare
   ▶ *Brachypalpus laphriformis* p. 282 – pictured right

   Underside of thorax hairy
   ▶ *Chalcosyrphus eunotus* p. 284

SAWFLY MIMICS

4 from 1e

a) **Legs partly yellow**
   ▶ *Xylota* pp. 286–289
   ▶ *Chalcosyrphus nemorum* p. 284
   See account for further information on how to distinguish *Chalcosyrphus nemorum* from some extremely similar *Xylota*.

b) **Legs black**
   ▶ *Brachypalpoides lentus* p. 280
   ▶ *Chalcosyrphus piger* (N/I)
   Looks very similar but with more enlarged hind femora.

277

XYLOTINI: BLERA

Typical Xylotini wing (*Xylota*)

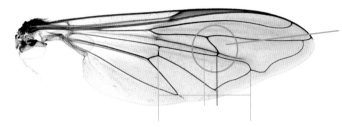

The Xylotini are enormously varied in form but a consistent character is the position of the R–M vein, which reaches the upper margin of the discal cell at the middle (*Syritta, p. 290*) or a point beyond the middle of this cell.

from 1a
p. 276

Blera

1 British species (illustrated)

The single species in this genus is very distinctive and extremely rare. The larvae inhabit rotting cavities in pine stumps. Its natural habitat is probably shattered stumps of windblown pines that had been weakened by the fungus *Phaeolus schweinitzi*. Recent studies have shown that suitable habitat can be created by making holes in pine stumps with a chainsaw.

LC
CR

Blera fallax Pine Hoverfly

GB: Rare [no trend data]
Ir: Not recorded

Wing length: 8·0–9·5 mm

Identification: A very distinctive hoverfly. The combination of a black thorax and base to the abdomen contrasting with orange-red terminal segments is unique among British hoverflies.

Similar species: None.

Observation tips: Only two small natural populations are known, both in stumps in pine plantations along the River Spey in central Scotland. There are also three translocation sites populated from captive-bred larvae and adults. It probably takes larvae several years to develop and, as the population is very small, adults are rarely seen. There are occasional sightings of adults at Rowan and Raspberry.

J F M A M J J A S O N D

A conservation success story?

Blera fallax will mate and lay eggs in captivity but females require pollen to provide additional protein to aid egg maturation; Rowan is the most favoured flower. Captive-bred larvae and adults have been translocated successfully to establish viable new populations using artificial rot-holes (RIGHT) that are good substitutes for natural ones. There is, however, concern that a genetic bottleneck arose from severe population depletion about 150 years ago, which the introduction of new genetic lines from Scandinavia may be required to address.

Blera fallax larvae in an artificial rot-hole

XYLOTINI: *Blera*

♂

Blera fallax × **5**

♀

XYLOTINI: BRACHYPALPOIDES | CALIPROBOLA

Brachypalpoides
from 4b
p. 277

1 British and Irish species (illustrated)

A very distinctive hoverfly which is a denizen of old woodlands, especially oakwoods. It is closely related to *Xylota* (pp. 286–289) and the adults behave in a similar way – running over leaves as they feed on honeydew and the pollen grains trapped in this sticky covering. Adults may be mimicking sawflies such as *Tenthredo* and *Macrophya* and/or spider-hunting wasps. Larvae live in the soft, wet, rotten wood of large, dead tree roots.

LC *Brachypalpoides lentus*

GB: Frequent ▽
Ir: Scarce

Wing length: 10–12 mm

Identification: The combination of black body, black legs and blood-red markings on the base of the abdomen is unique.

Similar species: Confusion is likely only with *Xylota segnis* (*p. 288*) and *X. tarda* (N/I), both of which have partly yellow legs and orange-red (rather than blood-red) abdominal markings. *Chalcosyrphus piger* (N/I – see *p. 284*), recorded from Suffolk in 2021, has red on the abdomen extending to T4, and much more enlarged hind femora than *Brachypalpoides lentus*.

Observation tips: Females can often be found flying low down around the roots of oaks and other large deciduous trees. Otherwise, adults are usually observed as they feed on the surface of sunlit leaves in woodland rides and clearings. It is known from fewer than 40 records in Ireland, where it seems to have a highly localised distribution

J F M A M J J A S O N D

Caliprobola
from 1a
p. 276

1 British species (illustrated)

This is one of the largest and most spectacular of Britain's hoverflies. The larvae live in soft, wet, decaying wood with a porridge-like consistency in the dead roots of old Beech trees.

LC NT *Caliprobola speciosa*

GB: Rare [no trend data]
Ir: Not recorded

Wing length: 11·0–12·5 mm

Identification: Readily identified by its unusual metallic greenish-black colour, orange legs and relatively long, narrow, orange-brown tinged wings.

Similar species: None.

Observation tips: A scarce species confined to the New Forest, Hampshire and Windsor Great Park, Surrey and Berkshire. Recent work in the New Forest suggests that it has become harder to find. Although a strong flier, this species has not shown any sign of dispersing from its two known localities. This makes it potentially very vulnerable to any change, such as drought, which might lead to the loss of ancient Beech trees.

J F M A M J J A S O N D

XYLOTINI: *Brachypalpoides* | *Caliprobola*

♂ *Brachypalpoides lentus* × 5

Models that this species and others with a similar abdominal pattern, such as *Xylota segnis* (p 288) may mimic: the sawfly *Macrophya annulata* (TOP INSET); a spider-hunting wasp *Priocnemis perturbator* (BOTTOM INSET).

♀ *Caliprobola speciosa* × 5

A large Beech stump supporting *C. speciosa*.

XYLOTINI: BRACHYPALPUS See also *Identifying wasp and bee mimics pp. 62–65*

from 3b **Brachypalpus** 1 British and Irish species (illustrated)

p. 277 A woodland species that is quite a convincing bee mimic. It is most frequently likened to an *Osmia* bee as males often fly around dead timber in a similar manner. The larvae are found in wet rot-holes in large, old, deciduous trees, especially oaks.

 Brachypalpus laphriformis

GB: Local =
Ir: Rare

Wing length: 8·5–10·75 mm

Identification: A bee mimic with enlarged hind femora, which are arched in males. The underside of the thorax, between the middle and hind legs, is bare (see *opposite*).

Similar species: *Chalcosyrphus eunotus* (p. 284), which has a hairy underside to the thorax (see *opposite*). and *Criorhina asilica* (p. 292), which does not have enlarged hind femora and has the distinctive *Criorhina* face (see *page 293*).

Observation tips: Occurs in the vicinity of old trees with wet rot-holes, especially oaks. It is largely a western species, occurring as far north as the Lake District but also with two recent records from Scotland that may represent range expansion or possibly more detailed recording effort. It is known from just four pre-2000 records in Ireland.

J F M A M J J A S O N D

Osmia bee – in flight, *Brachypalpus* mimics species from this genus..

XYLOTINI: *Brachypalpus*

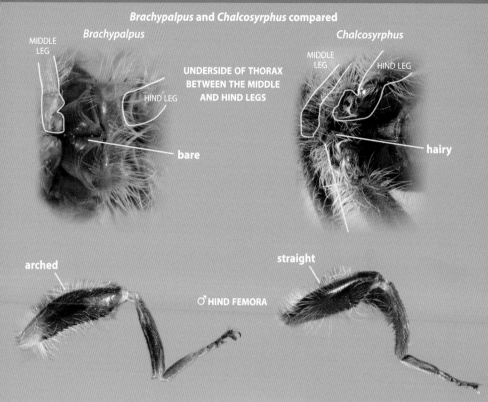

***Brachypalpus* and *Chalcosyrphus* compared**

♂ *Brachypalpus laphriformis* × **7**

XYLOTINI: *CHALCOSYRPHUS* See also *Identifying wasp and bee mimics pp. 62–65*

from 3b, 4
p. 277

Chalcosyrphus 3 British and Irish species (2 illustrated)

This is a very variable genus in Europe and the three British species are rather different from one another. **Unlike related genera, such as *Xylota* (pp. 286–289), the underside of the thorax, between the middle and hind legs, is hairy.** Larvae live in decaying sap under the bark of waterlogged timber and, as a consequence, adults are generally found in or near wet woodland.

LC *Chalcosyrphus nemorum*

LC

GB: Frequent =
Ir: Local

Wing length: 6·5–8·25 mm

Identification: A medium-sized xylotine with markings on abdomen segments T2 and T3 and hairs on the underside of the thorax. The hind femora are somewhat swollen when viewed in profile for comparative purposes (see *opposite*). The hind legs are black, with a small yellow section at the joint with the femur.

Similar species: *Xylota abiens* (no species account, but see image *opposite*), which appears a little more elongate (difficult to assess). However, no other xylotine has a hairy underside to the thorax. Smaller *X. jakutorum* (p. 286) are most likely to cause confusion but the hind tibia of that species has a substantial basal section.

Observation tips: A species of wet woodlands, particularly Alder carr. Adults are regular visitors to flowers, especially buttercups. In Britain, this is mainly a lowland species that is absent from higher elevations and especially tree-free zones. In Ireland, it has a somewhat patchy distribution and is mainly a southern species.

J F M A M J J A S O N D

LC *Chalcosyrphus eunotus*

LC

Nationally Scarce
GB: Scarce ▲
Ir: Not recorded

Wing length: 9·5–10·5 mm

Identification: This is a small honey-bee mimic with straight, enlarged hind femora and a hairy underside to the thorax, between the middle and hind legs.

Similar species: Most likely to be confused with *Brachypalpus laphriformis* [underside of thorax bare between middle and hind legs; hind femora enlarged (arched in male) – differences shown on *page 283*]. *Criorhina asilica* (p. 292) and older, paler examples of *C. floccosa* (p. 294) may also cause confusion but both those species lack enlarged hind femora and have a distinctive face-shape.

Observation tips: A highly localised species of 'dingle woodlands' (deep, wooded, stream gorges). The larvae live under the bark of partially submerged timber and this species is a flagship for a suite of invertebrates that are closely associated with log jams. This has become a rare habitat because log jams tend to get removed from streams and rivers by land managers to reduce the risk of flooding.

J F M A M J J A S O N D

Chalcosyrphus species not otherwise covered:

C. piger	Recorded from Suffolk in 2021. Quite similar to *Brachypalpoides lentus* (p. 280) but the red on the abdomen extends to segment T4 and the hind femora is much more enlarged.

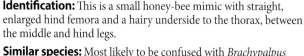

XYLOTINI: *Chalcosyrphus*

♂ *Chalcosyrphus nemorum* × 7

***Chalcosyrphus nemorum* and *Xylota abiens*
hind femora compared**

C. nemorum *X. abiens*

The hairy underside of the thorax distinguishes *Chalcosyrphus* from all other xylotines.

swollen towards apex

not swollen towards apex

Chalcosyrphus eunotus × 7

XYLOTINI: XYLOTA

from 4a **Xylota** — 7 British and Irish species (3 illustrated)
p.277 Rather elongate hoverflies that generally mimic sawflies or spider-hunting wasps (see *page 281*). Underside of thorax bare. Often seen running over leaves feeding on a mixture of pollen and honeydew rather than visiting flowers. The larvae live in decaying timber or wood debris.

Xylota with orange-red abdominal markings — 1/2

Xylota jakutorum

GB: Frequent ▲
Ir: Frequent

Wing length: 7·5–8·0 mm

Identification: Males have distinct orangish spots on abdomen segments T2 and T3, whilst in females the spots are often almost silvery. There are conspicuous spines along the whole length of the underside of the hind femur and the basal third of the hind tibia is conspicuously pale. Females have dusting on the frons that extends completely between the eyes (see image *below*)

Similar species: Most likely to be confused with *Xylota florum* (no species account, but see *opposite*), males of which have a longer abdomen due to the longer (and hence narrower-looking) segments and also have fewer and less conspicuous spines on the hind femur underside; female *X. florum* have two separated dust spots on the frons. Could also be confused with *X. abiens* (no species account, but see *page 285*) and *Chalcosyrphus nemorum* (*p. 285*), but in both cases the hind tibiae have little yellow at their base.

J F M A M J J A S O N D

Observation tips: Closely associated with decaying conifer timber, especially in recently felled areas, where it breeds in the rotting stumps. It can be abundant at buttercups and other flowers in rides and clearings in conifer plantations. Found mainly in northern and western Britain. In the 19th century it was regarded as a Caledonian pine wood species, but has spread southwards in pine plantations and may still be spreading south-eastwards. In Ireland it has a patchy distribution that reflects available habitat.

Frons of ♀ *X. jakutorum* showing complete band of dusting between the eyes.

Xylota species not otherwise covered:

X. abiens	NS	Most similar to *Chalcosyrphus nemorum* (see *page 285* for differences) and *X. jakutorum* (differences summarised *above*). A scarce species of southern England, but with scattered records north to Cumbria.
X. florum		Most similar to *X. jakutorum* (differences discussed *above*). Widespread in England and Wales, but less frequent than *X. jakutorum* and generally found in damp woodland.
X. tarda	NS	Difficult to separate from *X. segnis* (*p. 288*). A rather rare but widespread species associated with Aspen and other poplars. There is one pre-2000 Irish record.
X. xanthocnema	NS	Most similar to *X. sylvarum* (see *page 289* for differences). A scarce species of Wales and the southern half of England. Usually in or near ancient broadleaved woodland.

XYLOTINI: *Xylota*

♂ *Xylota jakutorum* × **7**

Selected *Xylota* compared

♀   ♂   **ABDOMENS**   ♀   ♂

X. jakutorum   *X. florum*

HIND FEMORA

SPINES ON UNDERSIDE conspicuous

SPINES ON UNDERSIDE fewer and less conspicuous than on *X. jakutorum*

XYLOTINI: XYLOTA

Xylota with orange-red abdominal markings 2/2

LC Xylota segnis
LC

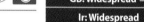

GB: Widespread =
Ir: Widespread

Wing length: 7·0–9·5 mm

Identification: The combination of elongate body-shape, orange-red markings on the abdomen and partly yellow legs makes this species distinct, with few possibilities for confusion. Some individuals have a trace of a black band at the base of abdomen segment T2, and these require microscopic examination of the spines on the hind leg for certain identification (see *opposite*).

Similar species: The most likely confusion is with the Nationally Scarce *Xylota tarda* (N/I), which is associated with Aspen: *X. tarda* has the orange marking on the abdomen divided by a black band at the base of segment T2 – but beware that some *X. segnis* can show a trace of this feature (see *above*). A cursory glance could result in confusion with three other species: *Tropidia scita* (*p. 290*), which has a femoral tooth; *Platycheirus granditarsus* (*p. 82*), males of which have modified front tarsi; and *Brachypalpoides lentus* (*p. 280*), which has black legs.

J F M A M J J A S O N D

Observation tips: Easily recognised by its behaviour: it scuttles back and forth, sweeping from side to side across leaves, collecting pollen and honeydew. In the south of Britain, it is rarely found at flowers as the less frequent rainfall allows honeydew to accumulate (trapping pollen in the process). In Scotland, where heavier rainfall washes away honeydew, it is a regular visitor to flowers such as Meadowsweet and Wild Angelica. Widespread and often abundant across Britain and Ireland.

♂ *X. segnis* in flight showing the orange-red abdominal markings.

Xylota with a golden 'tail' – X. sylvarum and X. xanthocnema

LC Xylota sylvarum
LC

GB: Widespread ∇
Ir: Frequent

Wing length: 7–12mm

Identification: One of two species that are distinctive among *Xylota* due to the patch of golden hairs at the end of the abdomen. *X. sylvarum* has hind tibiae that are partially black.

Similar species: Only the Nationally Scarce *Xylota xanthocnema* (no species account, but see *opposite*), which is scarce in southern Britain and not recorded from Ireland. It is smaller and daintier than *X. sylvarum* and best separated by the hind tibiae which are typically yellow (at most reddish-orange), never partially black.

Observation tips: This species is found in deciduous woodland, where it is a regular visitor to the flowers of white umbellifers such as Hemlock Water-dropwort. It also feeds on leaf surfaces in a manner much like *X. segnis*. Widespread in England and Wales but scarcer in Scotland; in Ireland it similarly has a mainly southerly distribution.

J F M A M J J A S O N D

XYLOTINI. *Xylota*

Xylota segnis × **7**

♂ **X. segnis**
HIND FEMUR underside with two rows of spines; TROCHANTER with long spurs

Xylota sylvarum × **7**

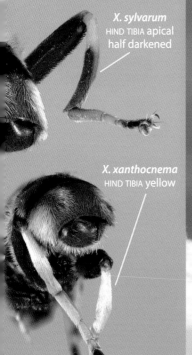

X. sylvarum
HIND TIBIA apical half darkened

X. xanthocnema
HIND TIBIA yellow

289

Syritta

from 1b p.276

1 British and Irish species (illustrated)

A small, dark, narrowly built hoverfly with greatly enlarged hind femora and the sides of the thorax heavily dusted ash-grey. Males are highly territorial; opponents will face up, forcing one another forwards and backwards until one gives up. Larvae live in wet, decaying matter such as compost, manure and silage, but not actually in water.

Syritta pipiens

GB: Widespread =
Ir: Widespread

Wing length: 4·25–7·0 mm

Identification: This species is unmistakable, provided you get close enough to see the greatly enlarged hind femora and ash-grey dusting on the thorax sides.

Similar species: There should be no confusion with other British and Irish species, although there is a very similar species, *Syritta flaviventris*, in southern Europe.

Observation tips: Often one of the commonest British species, which is absent only from more exposed and upland situations. It is similarly abundant in Ireland but inexplicably appears to be absent from parts of the west. Occasionally, there are mass occurrences that may be a result of immigration. It tends to be most abundant later in the season and into the autumn, and can be numerous on Ivy flowers.

J F M A M J J A S O N D

Tropidia

from 1b p.276

1 British and Irish species (illustrated)

A medium-sized, elongate hoverfly with conspicuous orange markings on the abdomen. It is immediately recognisable when examined closely as the hind femora have a pronounced triangular 'tooth'. Found in reedbeds, the larvae live in wet, decaying vegetation around reed bases.

Tropidia scita

GB: Frequent =
Ir: Frequent

Wing length: 5·5–8·25 mm

Identification: Has a distinctive pronounced triangular 'tooth' on the hind femur – only *Merodon equestris* (p. 248), a bumblebee mimic and hence very different looking, has anything similar.

Similar species: *Xylota segnis* (p. 288) and *X. tarda* (N/I – see page 286) are a somewhat similar shape and colour but lack the femoral 'tooth'.

Observation tips: Associated with beds of Common Reed, Bulrush and other tall emergent vegetation in flowing and still waters. Adults are flower visitors, but rarely stray far from breeding sites. In Britain it is commonest in river valleys and coastal marshes of southern and eastern England, although it does occur as far north as the Outer Hebrides. In Ireland it has a somewhat patchy distribution but seems to be most abundant towards the west coast.

J F M A M J J A S O N D

XYLOTINI: Syritta | Tropidia

♂ *Syritta pipiens* ×6

♂ *Tropidia scita* ×6

distinctive triangular 'tooth' on hind femora

Criorhina

from 2a, 3a
p. 277

4 British and Irish species (all illustrated)

Medium-sized to large bumblebee or honey-bee mimics with a **distinctive face-shape in which the mouth edge is elongated downwards.** The larvae live in wet, decaying subterranean timber. Females are often seen flying around the bases of tree stumps and large trees looking for egg-laying sites, and 'bees' behaving like this are therefore worth investigating.

LC *Criorhina asilica*

GB: Frequent =
Ir: Not recorded

Wing length: 9·5–11·25 mm

Identification: A good honey-bee mimic that is sexually dimorphic. It is somewhat elongate, with slightly inflated hind femora and greyish-orange markings on abdomen segments T2 and T3.

Similar species: Most likely to be confused with *Brachypalpus laphriformis* (p. 282) and *Chalcosyrphus eunotus* (p. 284), both of which have enlarged hind femora. In photographs in which the legs are difficult to discern and the wings cover the abdomen, it may be hard to determine this species' identity.

Observation tips: A woodland species which occurs widely in England and Wales. Males fly low and rapidly over low-growing vegetation in a manner similar to males of the solitary bee *Andrena scotica*. Both sexes are regular visitors to flowers such as Hawthorn and Dogwood. It seems to have become more frequent in recent years.

J F M A M J J A S O N D

LC *Criorhina ranunculi*

GB: Frequent ▲
Ir: Scarce

Wing length: 11·25–14·0 mm

Identification: Larger and darker than the other *Criorhina* species. There are several colour forms with red or white 'tails' that mimic different bumblebee species, but all forms have the expanded, arched hind femora that make *C. ranunculi* highly distinctive and unlikely to be confused with other bumblebee mimics.

Similar species: Other bumblebee mimics such as *Eristalis intricaria* (p. 228), *Merodon equestris* (p. 248) and *Volucella bombylans* (p. 270) – although none of these has the distinctive *Criorhina* face-shape (see *opposite*).

Observation tips: An early spring species which visits the flowers of Blackthorn, willows and Wild Cherry, often staying high up. Females can be found prospecting around the roots of trees, including birches on heathland and oaks in woodlands. Mainly a southern species in Britain, although there are scattered records north to the Moray Firth. In Ireland, it is known from just 13 records, although its early flight time may mean that it has been under-recorded.

J F M A M J J A S O N D

XYLOTINI: *Criorhina*

The distinctive face-shape of *Criorhina*

mouth edge elongated downwards

♂ *Criorhina asilica* × 4

♀ *Criorhina ranunculi* × 4 ♂

XYLOTINI: *CRIORHINA* See also *Identifying wasp and bee mimics pp. 62–65*

LC *Criorhina floccosa*

GB: Frequent =
Ir: Local

Wing length: 10–13 mm

Identification: A broad, furry species that is normally a warm-toned orange-brown, although darker hues can occur. The most distinctive feature is the tufts of pale hair on the sides of the thorax and at the base of the abdomen, although these hair tufts are not always obvious. The antennae are usually dark brownish-black but can be paler (beware confusion with *Criorhina berberina* var. *oxyacanthae* – see *below*).

Observation tips: This is a woodland species which frequently basks on sunlit leaves and visits a range of flowers such as Hawthorn and Dogwood. It often shows interest in the bases of a range of deciduous trees and the hollow trunks of ancient Ash pollards. *C. floccosa* is widely distributed across Britain but becomes scarcer in Scotland. It has a much sparser distribution in Ireland, where it appears to be mainly coastal and southern.

J F M A M J J A S O N D

Similar species: *Criorhina floccosa* and *C. berberina* var. *oxyacanthae* are readily confused but can usually be separated by antennae colour (dark in the former, normally orange in the latter) – but intermediates do occur. Often confused with *Merodon equestris* (p. 248) but that species has a characteristic loop in vein R$_{4+5}$. *Volucella bombylans* (p. 270) looks similar but has plumose aristae and dark markings on the wings. Typical *C. berberina* could be confused with *Pocota personata* (p. 296) if the size of the head relative to the body is not obvious.

LC *Criorhina berberina*

GB: Frequent ▽
Ir: Local

Wing length: 8–12 mm

Identification: A bumblebee mimic with two colour forms. The typical form is dark with a buff stripe across the front of the thorax and a pale-haired 'tail'. The other form, var. *oxyacanthae*, is entirely buff-coloured with the fur on the sides of the thorax and abdomen relatively uniform in length and coloration. The antennae are usually orange but browner shades do occur.

Observation tips: A woodland and hedgerow species that occurs widely across Britain and much of southern Ireland. The adults are regular visitors to a wide range of flowers including Hogweed.

J F M A M J J A S O N D

Both *Criorhina floccosa* and *C. berberina* var. *oxyacanthae* mimic the carder bee *Bombus pascuorum* (shown here).

XYLOTINI: *Criorhina*

♂ *Criorhina floccosa* × 5

tufts of pale hair at base of abdomen

♀ *Criorhina berberina* × 5

Var. *oxyacanthae* (ABOVE) is very different from the typical form (BELOW).

XYLOTINI: POCOTA   See also *Identifying wasp and bee mimics pp. 62–65*

from 3b
p. 277

Pocota

1 British species (illustrated)

A very convincing bumblebee mimic, the adults even emitting a loud buzz that sounds like an angry bee and, when disturbed, lifting a hind leg in the same manner as a bumblebee. **It is distinct from other hoverflies because it has a disproportionately small head in comparison to the body.** The larvae live in wet rot-holes in Beech and it has been suggested that they prefer holes which are high up.

LC *Pocota personata*

Nationally Scarce	
GB: Scarce =	
Ir: Not recorded	

Wing length: 11–13 mm

Identification: Highly distinctive bumblebee mimic which is unmistakable once the small head is seen. The yellow hairs on the thorax and abdomen are strikingly lemon-yellow, contributing to its unique appearance.

Similar species: If the small head is not noticed, may be misidentified as *Criorhina berberina* (*p. 294*) or *Volucella bombylans* (*p. 270*).

Observation tips: Although there are only scattered records from Wales and England north to County Durham, this species may not be as scarce as this suggests, as the very short flight-period (peaking late April to early May) means that it can be easily overlooked. Searching for larvae in rot-holes may be a more efficient way of finding this species. Both sexes occasionally visit flowers.

J F M A M J J A S O N D

♂ *Pocota personata* × 5

A mating pair.

XYLOTINI. *Pocota*

♂ *Pocota personata* × 5

head noticeably small in relation to body

MICRODONTINAE: *MICRODON*

from 4 p.56

Microdon

4 British and Irish species (all illustrated)

Dumpy hoverflies with short, pointed wings and long, forward-pointing antennae. The eyes are separated in both sexes. The larvae live in ants' nests where they feed on the eggs and brood of the ants. Adults are not strong fliers and tend to be found close to breeding sites. They are hard to find and all species are most easily located by looking for larvae and puparia in ants' nests.

NT
LC

Microdon analis

Nationally Scarce

GB: Local =
Ir: Rare

Wing length: 6·75–8·25 mm

Identification: This species has a dark scutellum and the hairs on the disc of the thorax are pale.

Similar species: *Microdon* are unlike any other hoverfly. Within the genus, however, the species that can be separated are superficially similar and require careful examination (see table opposite). *M. analis* is most likely to be confused with *M. devius* (p. 300) but that species has a zone of partly black hairs on the thorax between the wing bases.

Observation tips: Larvae live in the nests of black ants of the *Lasius niger/platythorax* complex on heathlands. Adults rarely fly far and do not appear to be flower visitors. The distribution is disjunct: they occur on the heaths of central southern England and in Caledonian pine woods in the north of Scotland. Known from just 11 records in Ireland, but since this is a difficult species to find it may prove to be commoner than records suggest.

J F M A M J J A S O N D

Microdon – armoured intruders

Microdon larvae feed on ant larvae and pupae within ants' nests, and are superbly adapted to this life. The larvae recall a military tank: they are hemispherical with a distinctly sculptured upper surface and a dense fringe of hairs that helps prevent attacks by ants. The head is usually concealed and only protrudes through the fringe of protective hairs to grab prey. The larva's shape also enables it to cling tightly to a surface, making it difficult for an ant to penetrate its defences. The larvae absorb the ants' pheromones, making them chemically indistinguishable from the ants themselves. This capability for biochemical camouflage is passed to the next hoverfly generation via the egg, which is why the first instar larvae are able to enter a host nest unchallenged.

The relationship between ants and *Microdon* colonisation is complex and is not the same in all species. Recent studies of the biology of *Microdon mutabilis* and *M. myrmicae* has highlighted some differences. In *M. mutabilis*, the farther a female was moved from the original ant colony from which she emerged, the less likely were her offspring to be able to enter an ants' nest. However, *M. myrmicae* did not seem to show this relationship. These biochemical interactions between *Microdon* larvae and ants open an intriguing window onto the issues that may affect the conservation of our rarer *Microdon* species.

So far, *M. mutabilis* and *M. myrmicae* can only be distinguished by features of their larvae and puparia. It is possible that there are also two or more species within *M. analis* that may similarly be distinguishable only at the immature stages.

MICRODONTINAE: Microdon

Key differences between British and Irish Microdon species

SPECIES	SCUTELLUM	THORAX HAIRS
M. analis (opposite)	dark	all hairs on thorax **pale**
M. devius (p. 300)	dark	hairs on thorax between wing bases **partly black**
M. myrmicae (p. 300)	red-brown	adults cannot be separated; they are distinguished by features of the mature larvae/puparia
M. mutabilis (p. 300)	red-brown	

♀ *Microdon analis* ×6

THORAX with pale hairs

M. analis puparium with ant

Collection of *M. analis* puparia under bark

Microdon devius

NT
NT

Wing length: 6·25–9·25 mm

Identification: The largest of the British *Microdon* species. Separated from the others by its dark scutellum and the presence of a zone of partly black hairs on the thorax between the wing bases.

Similar species: *Microdon* are unlike any other hoverfly. Within the genus, however, the species that can be separated are superficially similar and require careful examination (see table on page 299). Most likely to be confused with *M. analis* (p. 298), which has pale hairs on the top of the thorax.

Observation tips: Larvae live in the nests of the Yellow Meadow Ant. The majority of colonies are on chalk downland but there are also outlying colonies in north Wales and Norfolk. Adults are usually found basking on foliage, but they occasionally visit flowers. Although not fully understood, this species appears to have very specific habitat requirements with respect to the location of anthills. It is also susceptible to changes in grassland management: for example, a switch from grazing to mowing has been known to wipe out a colony.

J F M A M J J A S O N D

Microdon myrmicae¹ / mutabilis²

LC
VU
DD

Wing length: 6·5–9·0 mm

Identification: The reddish-brown scutellum distinguishes these two species from other *Microdon*, but the **adults cannot be separated**. Although geographical location may provide a good clue, identification can only be confirmed by features of the mature larvae or puparium.

Similar species: *Microdon* are unlike any other hoverfly. Within the genus, however, the species that can be separated are superficially similar and require careful examination (see table on page 299). Only likely to be confused with *M. analis* (p. 298) or *M. devius*, both of which have a dark scutellum.

Observation tips: These species have very different ant associations: *M. myrmicae* larvae live in the nests of the ant *Myrmica scabrinodis* in tussocks in wet pastures and in bogs, where as those of *M. mutabilis* live in the nests of the ant *Formica lemani* under stones in places such as limestone pavements. Both are primarily 'west coast species'. *M. myrmicae* appears to be far more widely distributed, being locally common in southern and western Britain, whereas *M. mutabilis* has been found mainly in on the limestone of north Lancashire and on Mull on the west coast of Scotland. It is unsafe to assume that habitat is indicative of the species because in some places suitable habitat is juxtaposed (*e.g.* in the Arnside area of north Lancashire). Searching for larvae of *M. myrmicae* may show it to have a far wider distribution. Although both species are reported from Ireland, records are very sparse and they are likely to have been under-recorded.

J F M A M J J A S O N D

MICRODONTINAE: *Microdon*

THORAX zone of partly black hairs between the wing bases

a mating pair

♀ *Microdon devius* × 6

M. devius
scutellum **dark**

M. myrmicae/mutabilis
scutellum **reddish-brown**

♀ *Microdon myrmicae/mutabilis* × 6

List of British and Irish hoverflies

The following table lists all 285 species of hoverfly recorded in Britain and Ireland up to the end of 2023, plus two additional 'species' of uncertain taxonomic status. It also shows the classification into subfamilies and tribes. The taxonomy used follows *British Hoverflies* (second edition) by Stubbs and Falk (2002) and the nomenclature follows Chandler (2022). The tribes in particular are useful in referring to features shared by groups of genera (see *pages 53–61*).

The '**Frequency**' columns are based on the number of Ordnance Survey National 10 km Grid Squares for Great Britain (**GB**) and Ireland (**Ir**) (see *page 71* for details).

The 'Trend' (**TR**) column shows changes in the frequency with which the species has been recorded in Great Britain according to records submitted to the Hoverfly Recording Scheme from 1980 to 2022.

▲ Frequency of recording has increased significantly
▼ Frequency of recording has decreased significantly
= No significant change in recording frequency
– Insufficient data for a trend to be calculated

The columns with red headers show the International Union for Conservation of Nature (IUCN) Red List status for all species recorded in Great Britain (**GB**) and, separately, Europe (**Eu**); at present there is no Red List of hoverflies for Ireland. For GB, this is based on Ball & Morris (2014); for Europe, it is based on Vujić *et al*. (2022). Red Lists provide an assessment of the threat of extinction of species based on internationally agreed criteria developed by the IUCN (IUCN, 2012). These criteria include rates of decline, geographic range and population size. The IUCN maintains and regularly updates global inventories of threatened species. The same IUCN Red List guidelines are used to produce Red Lists for different groups of animals and plants at a regional or national level. These provide an indicator of the health of our biodiversity and inform conservation action.

Species assessed as being threatened with extinction (Red Listed) are categorised (and coded in this book) as **Critically Endangered** CR, **Endangered** EN or **Vulnerable** VU.
In addition, species may be categorised as **Near Threatened** NT if they are close to qualifying for, or likely to qualify for, a threatened category in the near future. The remaining are considered to be of **Least Concern** LC. Species for which the data are considered inadequate to make an assessment may be categorised as **Data Deficient** DD. The **Not Evaluated** NE category is used for recently discovered species or those that have occurred only as a vagrant. Species that are taxonomically contentious or uncertain on European lists are coded as **Not Recognised** NR.

For Great Britain, the British Rarity Status (**BRS**) is shown, with codes used as follows:
NS   Nationally Scarce (which is applied to species that are expected to occur in 16–100 10-km grid squares).
BAP Included in the UK Biodiversity Action Plan (see *page 322*).
vag Believed to occur only as a vagrant and does not have an established population.

The 'Difficulty of Identification' (**ID**) and 'Photo Identification' (**Photo**) icons are as used in the species accounts (see *page 70*).

Page numbers are coded as follows:
Bold black text for species with full accounts.
Bold grey text for species referenced in accounts with accompanying images.
Light black text for species mentioned in accounts or in tables, but not illustrated.

LIST OF BRITISH AND IRISH HOVERFLIES

Name	ID	Photo	TR	Frequency (GB)	GB	BRS	Frequency (Ir)	Eu	Page
Sub-family: SYRPHINAE									
Tribe: BACCHINI									
Baccha									
B. elongata (Fabricius, 1775)	👁	📷	▼	Widespread	LC		Widespread	LC	74
Melanostoma									
* M. dubium (Zetterstedt, 1838)	ℯ	✗	▼	Rare	LC	NS		LC	95
M. mellinum (Linnaeus, 1758)	🔎	📷	=	Widespread	LC		Widespread	LC	94
M. scalare (Fabricius, 1794)	🔎	📷	▲	Widespread	LC		Widespread	LC	94
Platycheirus									
P. albimanus (Fabricius, 1781)	🔎	❓	=	Widespread	LC		Widespread	LC	84
P. ambiguus (Fallén, 1817)	ℯ	❓	▼	Frequent	LC		Local	LC	84
P. amplus Curran, 1927	ℯ	✗	–	Rare	NT		Rare	LC	80
P. angustatus (Zetterstedt, 1843)	ℯ	❓	=	Widespread	LC		Frequent	LC	90
P. aurolateralis Stubbs, 2002	ℯ	✗	–	Scarce	LC		Rare	LC	80
P. clypeatus (Meigen, 1822)	ℯ	❓	=	Widespread	LC		Widespread	LC	90
P. discimanus Loew, 1871	ℯ	❓	=	Local	LC	NS	Rare	LC	80
P. europaeus Goeldlin, Maibach & Speight, 1990	ℯ	✗	–	Local	LC			LC	80
P. fulviventris (Macquart, 1827–8)	ℯ	❓	▼	Frequent	LC		Frequent	LC	92
P. granditarsus (Forster, 1771)	👁	📷	=	Widespread	LC		Widespread	NT	82
P. immarginatus (Zetterstedt, 1849)	ℯ	✗	▼	Local	LC	NS	Scarce	NT	80
P. manicatus (Meigen, 1822)	🔎	📷	▼	Widespread	LC		Frequent	LC	86
P. melanopsis Loew, 1856	ℯ	❓	▼	Rare	NT			LC	80
P. nielseni Vockeroth, 1990	ℯ	✗	–	Frequent	LC		Local	LC	80
P. occultus Goeldlin, Maibach & Speight, 1990	ℯ	✗	–	Frequent	LC		Frequent	LC	80
P. peltatus (Meigen, 1822)	ℯ	❓	=	Widespread	LC		Frequent	LC	88
P. perpallidus Verrall, 1901	ℯ	✗	▼	Scarce	LC	NS	Local	LC	80
P. podagratus (Zetterstedt, 1838)	ℯ	✗	=	Local	LC		Rare	LC	80
P. ramsarensis Goeldlin, Maibach & Speight, 1990	ℯ	✗	–	Frequent	LC		Scarce	LC	80
P. rosarum (Fabricius, 1787)	👁	📷	=	Widespread	LC		Frequent	LC	82
P. scambus (Staeger, 1843)	ℯ	✗	–	Frequent	LC		Local	LC	80
P. scutatus (Meigen, 1822)	ℯ	❓	▼	Widespread	LC		Frequent	LC	88
P. splendidus Rotheray, 1998	ℯ	❓	–	Frequent	LC		Rare	LC	80
P. sticticus (Meigen, 1822)	ℯ	✗	–	Scarce	LC	NS	Rare	LC	80
P. tarsalis (Schummel, 1836)	🔎	❓	▼	Frequent	LC			LC	86
Xanthandrus									
X. comtus (Harris, 1780)	👁	📷	=	Frequent	LC		Local	LC	92
Tribe: PARAGINI									
Paragus									
P. albifrons (Fallén, 1817)	ℯ	❓	–	Rare	CR			EN	96
P. constrictus Simic, 1986	ℯ	✗					Rare	EN	96
P. haemorrhous Meigen, 1822	ℯ	✗	▲	Widespread	LC		Frequent	LC	96
P. quadrifasciatus Meigen, 1822	ℯ	❓	–	Rare	NE			LC	96
P. tibialis (Fallén, 1817)	ℯ	✗	=	Rare	NT			LC	96

---

*Recent DNA analysis sinks *Melanostoma dubium* as a species and erects two new ones: *M. mellarium* and *M. certuum*, both of which are likely to be added to the British list in due course. The European Red List recognises both these species and categorises each as LC.

LIST OF BRITISH AND IRISH HOVERFLIES

Name	ID	Photo	TR	Frequency (GB)	GB	BRS	Frequency (Ir)	Eu	Page
Tribe: SYRPHINI									
Chrysotoxum									
C. arcuatum (Linnaeus, 1758)	🔍	📷	▼	Frequent	LC		Frequent	LC	104
C. bicinctum (Linnaeus, 1758)	👁	📷	▲	Widespread	LC		Widespread	LC	102
C. cautum (Harris, 1776)	🔍	📷	▼	Frequent	LC		Rare	LC	104
C. elegans Loew, 1841	⚢	❓	=	Local	LC	NS		NT	106
C. festivum (Linnaeus, 1758)	👁	📷	=	Widespread	LC		Local	LC	102
C. octomaculatum Curtis, 1837	⚢	✖	–	Rare	EN	BAP		NT	106
C. vernale Loew, 1841	⚢	❓	–	Rare	EN			LC	106
C. verralli Collin, 1940	⚢	❓	=	Frequent	LC			LC	106
Dasysyrphus									
D. albostriatus (Fallén, 1817)	👁	📷	▼	Widespread	LC		Frequent	LC	120
D. friuliensis van der Goot, 1960	⚢	✖	–	Rare	LC			LC	122
D. hilaris (Zetterstedt, 1843)	⚢	✖	–	Rare	LC		Rare	LC	122
D. neovenustus Soszyński et al., 2013	⚢	✖	–	Rare	NE		Rare	LC	122
D. pauxillus (Williston, 1887)	⚢	✖	–	Rare	LC			LC	122
D. pinastri (De Geer, 1776)	⚢	✖	▼	Frequent	LC		Rare	LC	122
D. tricinctus (Fallén, 1817)	👁	📷	▼	Widespread	LC		Local	LC	120
D. venustus (Meigen, 1822)	🔍	✖	▼	Widespread	LC		Local	LC	122
Didea									
D. alneti (Fallén, 1817)	⚢	❓	–	Rare	LC	vag	Rare	LC	124
D. fasciata Macquart, 1834	⚢	❓	▼	Frequent	LC		Frequent	LC	124
D. intermedia Loew, 1854	⚢	✖	=	Scarce	LC	NS		LC	124
Doros									
D. profuges (Harris, 1780)	👁	📷	=	Scarce	NT	BAP	Rare	LC	108
Epistrophe									
E. diaphana (Zetterstedt, 1843)	⚢	📷	=	Frequent	LC			LC	138
E. eligans (Harris, 1780)	👁	📷	▲	Widespread	LC		Frequent	LC	136
E. flava Doczkal & Schmid, 1994	⚢	✖	–	Rare	NE	vag		LC	136
E. grossulariae (Meigen, 1822)	🔍	📷	▼	Widespread	LC		Frequent	LC	138
E. melanostoma (Zetterstedt, 1943)	⚢	📷	=	Local	LC	NS		LC	140
E. nitidicollis (Meigen, 1822)	⚢	📷	=	Frequent	LC		Scarce	LC	140
E. ochrostoma (Zetterstedt, 1849)	⚢	✖	=	Rare	DD			LC	136
Episyrphus									
E. balteatus (Degeer, 1776)	👁	📷	▲	Widespread	LC		Widespread	LC	162
Eriozona									
E. syrphoides (Fallén, 1817)	🔍	📷	▼	Frequent	LC	NS	Local	LC	116
Eupeodes									
E. bucculatus (Rondani, 1857)	⚢	✖	=	Local	LC		Local	LC	146
E. corollae (Fabricius, 1794)	👁	📷	=	Widespread	LC		Widespread	LC	148
E. goeldlini Mazánek, Láska & Bicík, 1999	⚢	✖	–	Rare	LC		Rare	LC	146
E. lapponicus (Zetterstedt, 1838)	🔍	❓	–	Local	LC	vag	Rare	LC	150
E. latifasciatus (Macquart, 1829)	🔍	📷	▼	Widespread	LC		Frequent	LC	148
E. lundbecki (Soot-Ryen, 1946)	⚢	✖	–	Rare	LC	vag		LC	146
E. luniger (Meigen, 1822)	🔍	📷	=	Widespread	LC		Frequent	LC	147
E. nielseni Dušek & Láska 1976	⚢	✖	▼	Scarce	LC	NS		LC	146
E. nitens (Zetterstedt, 1843)	⚢	✖	=	Scarce	LC	NS		LC	146

LIST OF BRITISH AND IRISH HOVERFLIES

Name	ID	Photo	TR	Frequency (GB)	GB	BRS	Frequency (Ir)	Eu	Page
Leucozona									
L. glaucia (Linnaeus, 1758)	👁	📷	▼	Widespread	LC		Widespread	LC	118
L. laternaria (Muller, 1776)	👁	📷	=	Widespread	LC		Frequent	LC	118
L. lucorum (Linnaeus, 17581)	👁	📷	=	Widespread	LC		Widespread	LC	116
Megasyrphus									
M. erraticus (Linnaeus, 1758)	🔍	📷	▼	Local	LC	NS	Rare	LC	131
Melangyna									
M. arctica (Zetterstedt, 1838)	⚥	✗	▼	Local	LC		Local	LC	158
M. barbifrons (Fallén, 1817)	⚥	✗	=	Rare	NT			LC	158
M. cincta (Fallén, 1817)	🔍	📷	▼	Frequent	LC		Local	LC	158
M. compositarum (Verrall, 1873)	🔍	📷	=	Widespread	LC		Rare	LC	156
M. ericarum (Collin, 1946)	⚥	✗	–	Rare	VU			NT	158
M. lasiophthalma (Zetterstedt, 1843)	⚥	📷	=	Frequent	LC		Frequent	LC	154
M. quadrimaculata (Verrall, 1873)	⚥	📷	▼	Local	LC		Rare	LC	158
M. umbellatarum (Fabricius, 1794)	⚥	📷	=	Widespread	LC		Local	LC	156
Meligramma									
M. euchromum (Kowarz, 1885)	🔍	📷	▼	Local	LC	NS		LC	150
M. guttatum (Fallén, 1817)	🔍	📷	▼	Local	LC	NS	Rare	LC	152
M. trianguliferum (Zetterstedt, 1843)	🔍	📷	▼	Frequent	LC		Rare	LC	152
Meliscaeva									
M. auricollis (Meigen, 1822)	🔍	📷	=	Widespread	LC		Widespread	LC	160
M. cinctella (Zetterstedt, 1843)	👁	📷	▼	Widespread	LC		Widespread	LC	160
Parasyrphus									
P. annulatus (Zetterstedt, 1838)	⚥	✗	▼	Scarce	LC		Rare	LC	142
P. lineola (Zetterstedt, 1843)	⚥	✗	▼	Local	LC		Rare	LC	142
P. malinellus (Collin, 1952)	⚥	✗	▼	Local	LC		Scarce	LC	142
P. nigritarsis (Zetterstedt, 1843)	⚥	📷	=	Local	LC	NS	Scarce	LC	144
P. punctulatus (Verrall, 1873)	🔍	📷	=	Widespread	LC		Frequent	LC	142
P. vittiger (Zetterstedt, 1843)	🔍	📷	▼	Local	LC		Rare	LC	144
Scaeva									
S. albomaculata (Macquart, 1842)	⚥	📷	–	Rare	NE	vag		LC	126
S. dignota (Rondani, 1875)	⚥	📷	–	Rare	NE	vag	Rare	LC	126
S. mecogramma (Bigot, 1860)	⚥	📷	–	Rare	NE	vag		LC	126
S. pyrastri (Linnaeus, 1758)	👁	📷	=	Widespread	LC		Widespread	LC	126
S. selenitica (Meigen, 1822)	🔍	📷	=	Frequent	LC		Local	LC	126
Sphaerophoria									
S. bankowskae Goeldlin, 1974	⚥	✗	–	Rare	DD			LC	112
S. batava Goeldlin, 1974	⚥	✗	=	Local	LC		Rare	LC	112
S. fatarum Goeldlin, 1989	⚥	✗	=	Local	LC		Local	NT	112
S. interrupta (Fabricius, 1805)	⚥	✗	=	Widespread	LC		Frequent	LC	112
S. loewi Zetterstedt, 1843	⚥	📷	–	Rare	NT		Rare	NT	112
S. philanthus (Meigen, 1822)	⚥	✗	=	Frequent	LC		Local	LC	112
S. potentillae Claussen, 1984	⚥	✗	–	Rare	VU			VU	112
S. rueppellii (Wiedemann, 1830)	⚥	📷	▼	Frequent	LC		Rare	LC	114
S. scripta (Linnaeus, 1758)	👁	📷	▲	Widespread	LC		Frequent	LC	114
S. taeniata (Meigen, 1822)	⚥	✗	=	Frequent	LC			LC	112
S. virgata Goeldlin, 1974	⚥	✗	=	Scarce	LC	NS		NT	112
S. 'species B' sensu Stubbs 1991	⚥	✗	–	–	NE			NE	112

LIST OF BRITISH AND IRISH HOVERFLIES

Name	ID	Photo	TR	Frequency (GB)	GB	BRS	Frequency (Ir)	Eu	Page
Syrphus									
S. nitidifrons Becker, 1921	⚥	📷	–	Rare	LC			LC	132
S. rectus (Osten Sacken, 1875)	⚥	✗	–	Rare	LC		Rare	DD	132
S. ribesii (Linnaeus, 1758)	⚥	📷	=	Widespread	LC		Widespread	LC	132
S. torvus Osten Sacken, 1875	⚥	📷	=	Widespread	LC		Frequent	LC	134
S. vitripennis Meigen, 1822	⚥	✗	=	Widespread	LC		Frequent	LC	134
Xanthogramma									
X. citrofasciatum (Degeer, 1776)	👁	📷	=	Frequent	LC		Rare	LC	108
X. pedissequum (Harris, 1776)	👁	📷	▲	Widespread	LC			LC	110
X. stackelbergi Violovitsh, 1975	🔍	📷	–	Local	LC			LC	110

Sub-family: ERISTALINAE									
Tribe: CALLICERINI									
Callicera									
C. aurata (Rossi, 1790)	🔍	📷	=	Local	LC	NS		VU	166
C. rufa Schummel, 1841	👁	📷	=	Scarce	LC	NS		VU	164
C. spinolae Rondani, 1844	🔍	📷	–	Rare	VU	BAP		VU	166
Tribe: CHEILOSIINI									
Cheilosia									
C. ahenea von Roser, 1840	⚥	✗	–	Rare	VU		Local	DD	174
C. albipila Meigen, 1838	🔍	📷	▼	Frequent	LC		Local	LC	184
C. albitarsis Meigen, 1822	👁	✗	=	Widespread	LC		Frequent	LC	190
C. antiqua Meigen, 1822	⚥	✗	▼	Frequent	LC		Local	LC	174
C. barbata Loew, 1857	⚥	✗	▼	Scarce	LC	NS		LC	174
C. bergenstammi Becker, 1894	⚥	📷	=	Widespread	LC		Local	LC	188
C. caerulescens (Meigen, 1822)	🔍	📷	=	Frequent	LC			LC	176
C. carbonaria Egger, 1860	⚥	✗	=	Scarce	LC	NS		LC	175
C. chrysocoma (Meigen, 1822)	👁	📷	=	Local	LC	NS	Local	LC	186
C. cynocephala Loew, 1840	⚥	✗	=	Scarce	LC	NS		LC	175
C. fraterna (Meigen, 1830)	⚥	📷	▼	Frequent	LC			LC	186
C. griseiventris Loew, 1857	⚥	✗	–	Frequent	LC			NR	174
C. grossa (Fallén, 1817)	🔍	📷	=	Frequent	LC		Scarce	LC	184
C. illustrata (Harris, 1780)	👁	📷	=	Widespread	LC		Frequent	LC	176
C. impressa Loew, 1840	🔍	📷	▲	Widespread	LC		Rare	LC	190
C. lasiopa Kowarz, 1885	⚥	✗	▼	Frequent	LC			LC	174
C. latifrons (Zetterstedt, 1838)	⚥	✗	▼	Frequent	LC		Local	LC	174
C. longula (Zetterstedt, 1838)	⚥	✗	▼	Local	LC		Rare	LC	182
C. mutabilis (Fallén, 1817)	⚥	✗	▼	Scarce	LC	NS		LC	174
C. nebulosa Verrall, 1871	⚥	✗	–	Local	LC	NS	Rare	LC	175
C. nigripes (Meigen, 1822)	⚥	✗	=	Scarce	LC	NS		LC	174
C. pagana (Meigen, 1822)	🔍	📷	=	Widespread	LC		Frequent	LC	180
C. proxima (Zetterstedt, 1843)	🔍	✗	=	Widespread	LC			LC	188
C. psilophthalma Becker, 1894	⚥	✗	=	Scarce	DD		Rare	LC	175
C. pubera (Zetterstedt, 1838)	⚥	✗	–	Local	LC	NS	Rare	LC	174
C. ranunculi Doczkal, 2000	👁	📷	=	Frequent	LC			LC	190
C. sahlbergi Becker, 1894	⚥	✗	–	Rare	VU			DD	174
C. scutellata (Fallén, 1817)	🔍	📷	=	Frequent	LC		Scarce	LC	182

LIST OF BRITISH AND IRISH HOVERFLIES

Name	ID	Photo	TR	Frequency (GB)	GB	BRS	Frequency (Ir)	Eu	Page
Cheilosia (continued)									
C. semifasciata Becker, 1894	♂	✗	=	Rare	NT		Local	LC	175
C. soror (Zetterstedt, 1843)	♂	❓	▲	Frequent	LC			LC	180
C. urbana (Meigen, 1822)	♂	✗	▼	Local	LC			LC	175
C. uviformis (Becker, 1894)	♂	✗	–	Rare	DD		Rare	LC	175
C. variabilis (Panzer, 1798)	🔍	❓	=	Widespread	LC		Local	LC	178
C. velutina Loew, 1840	♂	✗	▼	Scarce	LC	NS	Rare	LC	175
C. vernalis (Fallén, 1817)	♂	✗	=	Widespread	LC		Local	LC	175
C. vicina (Zetterstedt, 1849)	♂	✗	▼	Local	LC		Rare	LC	174
C. vulpina (Meigen, 1822)	♂	✗	=	Frequent	LC			LC	178
C. 'species B'	♂	✗	–	Rare	DD			NE	175
Ferdinandea									
F. cuprea (Scopoli, 1763)	👁	📷	▲	Widespread	LC		Frequent	LC	192
F. ruficornis (Fabricius, 1775)	♂	❓	=	Scarce	LC	NS		LC	192
Portevinia									
P. maculata (Fallén, 1817)	👁	📷	=	Frequent	LC		Local	LC	192
Rhingia									
R. campestris Meigen, 1822	👁	📷	=	Widespread	LC		Widespread	LC	194
R. rostrata (Linnaeus, 1758)	🔍	📷	▲	Frequent	LC			LC	194
Tribe: CHRYSOGASTRINI									
Brachyopa									
B. bicolor (Fallén, 1817)	♂	❓	=	Local	LC	NS		LC	214
B. insensilis Collin, 1939	♂	✗	=	Local	LC		Rare	LC	212
B. pilosa Collin, 1939	♂	✗	▼	Local	LC	NS		LC	214
B. scutellaris RobineauDesvoidy, 1844	♂	✗	▲	Frequent	LC		Local	LC	212
Chrysogaster									
C. cemiteriorum (Linnaeus, 1758)	♂	❓	▼	Frequent	LC		Local	LC	206
C. solstitialis (Fallén, 1817)	🔍	📷	=	Widespread	LC		Frequent	LC	206
C. virescens Loew, 1854	♂	✗	▼	Frequent	LC		Rare	NT	206
Hammerschmidtia									
H. ferruginea (Fallén, 1817)	👁	📷	–	Rare	EN	BAP		LC	216
Lejogaster									
L. metallina (Fabricius, 1777)	♂	❓	▼	Widespread	LC		Frequent	LC	210
L. tarsata (Megerle in Meigen, 1822)	🔍	✗	=	Local	LC		Scarce	LC	210
Melanogaster									
M. aerosa (Loew, 1843)	♂	✗	▼	Local	LC		Local	LC	204
M. hirtella Loew, 1843	♂	✗	▼	Widespread	LC		Frequent	LC	204
Myolepta									
M. dubia (Fabricius, 1805)	🔍	❓	=	Local	LC	NS		LC	217
M. potens (Harris, 1780)	♂	✗	–	Rare	CR	BAP		LC	217
Neoascia									
N. geniculata (Meigen, 1822)	♂	✗	=	Local	LC		Local	LC	202
N. interrupta (Meigen, 1822)	♂	❓	=	Local	LC	NS		LC	202
N. meticulosa (Scopoli, 1763)	♂	✗	▼	Frequent	LC		Local	LC	200
N. obliqua Coe, 1940	♂	✗	▼	Local	LC		Rare	LC	202
N. podagrica (Fabricius, 1775)	♂	✗	▼	Widespread	LC		Widespread	LC	202
N. tenur (Harris, 1780)	♂	✗	▼	Widespread	LC		Frequent	LC	200

LIST OF BRITISH AND IRISH HOVERFLIES

Name	ID	Photo	TR	Frequency (GB)	GB	BRS	Frequency (Ir)	Eu	Page
Orthonevra									
O. brevicornis Loew, 1843	⚥	✗	▼	Local	LC			LC	209
O. geniculata Meigen, 1830	⚥	❓	▼	Local	LC		Local	LC	209
O. intermedia Lundbeck, 1916	⚥	✗	–	Rare	LC			LC	209
O. nobilis (Fallén, 1817)	🔍	❓	▼	Frequent	LC		Rare	LC	208
Riponnensia									
R. splendens (Meigen, 1822)	🔍	📷	▼	Widespread	LC		Frequent	LC	208
Sphegina									
S. clunipes (Fallén, 1816)	⚥	✗	=	Widespread	LC		Frequent	LC	198
S. elegans Schummel, 1843	⚥	✗	▼	Frequent	LC		Rare	LC	198
S. sibirica Stackelberg, 1953	🔍	❓	=	Frequent	LC		Rare	LC	200
S. verecunda Collin, 1937	⚥	✗	▲	Frequent	LC			LC	198
Tribe: ERISTALINI									
Anasimyia									
A. contracta Claussen & Torp, 1980	🔍	❓	=	Frequent	LC		Local	LC	240
A. interpuncta (Harris, 1776)	⚥	❓	=	Scarce	LC	NS		LC	241
A. lineata (Fabricius, 1787)	👁	📷	▼	Frequent	LC		Frequent	LC	240
A. lunulata (Meigen, 1822)	⚥	❓	=	Scarce	LC	NS	Local	LC	241
A. transfuga (Linnaeus, 1758)	⚥	❓	▼	Local	LC		Local	LC	241
Eristalinus									
E. aeneus (Scopoli, 1763)	🔍	📷	=	Frequent	LC		Local	LC	234
E. sepulchralis (Linnaeus, 1758)	🔍	📷	▼	Widespread	LC		Frequent	LC	234
Eristalis									
E. abusiva Collin, 1931	⚥	✗	▼	Frequent	LC		Frequent	LC	232
E. arbustorum (Linnaeus, 1758)	🔍	✗	=	Widespread	LC		Widespread	LC	232
E. cryptarum (Fabricius, 1794)	🔍	📷	–	Rare	CR	BAP	Rare	LC	228
E. horticola (Degeer, 1776)	🔍	📷	▼	Widespread	LC		Widespread	LC	230
E. intricaria (Linnaeus, 1758)	👁	📷	=	Widespread	LC		Widespread	LC	228
E. nemorum (Linnaeus, 1758)	🔍	📷	=	Widespread	LC		Frequent	LC	226
E. pertinax (Scopoli, 1763)	👁	📷	▲	Widespread	LC		Widespread	LC	224
E. rupium Fabricius, 1805	⚥	📷	▼	Frequent	LC			LC	230
E. similis (Fallén, 1817)	⚥	✗	–	Local	NE	vag?		LC	226
E. tenax (Linnaeus, 1758)	👁	📷	▲	Widespread	LC		Widespread	LC	224
Helophilus									
H. affinis Wahlberg, 1844	⚥	✗	–	Rare	NE	vag		LC	242
H. groenlandicus (Fabricius, 1780)	⚥	✗	–	Rare	DD			LC	242
H. hybridus Loew, 1846	🔍	📷	=	Widespread	LC		Widespread	LC	242
H. pendulus (Linnaeus, 1758)	🔍	📷	▲	Widespread	LC		Widespread	LC	242
H. trivittatus (Fabricius, 1805)	🔍	📷	▲	Widespread	LC		Frequent	LC	244
Lejops									
H. vittatus (Meigen, 1822)	🔍	📷	–	Rare	NT			VU	244
Mallota									
M. cimbiciformis (Fallén, 1817)	🔍	📷	=	Local	LC	NS		LC	236
Myathropa									
M. florea (Linnaeus, 1758)	👁	📷	▲	Widespread	LC		Widespread	LC	236
Parhelophilus									
P. consimilis (Malm, 1863)	⚥	✗	▼	Rare	LC	NS	Local	LC	238
P. frutetorum (Fabricius, 1775)	🔍	✗	▼	Frequent	LC			LC	238
P. versicolor (Fabricius, 1794)	🔍	✗	▼	Frequent	LC		Local	LC	238

Name	ID	Photo	TR	Frequency (GB)	GB	BRS	Frequency (Ir)	Eu	Page
Tribe: MERODONTINI									
Eumerus									
E. funeralis Meigen, 1822	⚥	✗	▼	Frequent	LC		Rare	LC	246
E. ornatus Meigen, 1822	⚥	📷	▼	Local	LC			LC	246
E. sabulonum (Fallén, 1817)	🔍	📷	=	Scarce	LC	NS		LC	248
E. sogdianus (Stackelberg, 1952)	⚥	✗	–	Rare	NE	vag		LC	246
E. strigatus (Fallén, 1817)	⚥	✗	▼	Frequent	LC		Local	LC	246
Merodon									
M. equestris (Fabricius, 1794)	👁	📷	=	Widespread	LC		Frequent	LC	248
Psilota									
P. anthracina Meigen, 1822	🔍	❓	=	Scarce	LC	NS		LC	250
Tribe: PELECOCERINI									
Pelecocera									
P. caledonicus Collin, 1940	⚥	✗	–	Rare	VU			LC	253
P. scaevoides (Fallén, 1817)	⚥	✗	▼	Rare	LC	NS		LC	252
P. tricincta Meigen, 1822	🔍	📷	▲	Scarce	LC	NS		LC	252
Tribe: PIPIZINI									
Heringia									
H. heringi (Zetterstedt, 1843)	⚥	✗	=	Local	LC		Rare	LC	260
H. senilis Sack, 1938	⚥	✗	–	Rare	LC			NR	261
Neocnemodon									
N. brevidens (Egger, 1865)	⚥	✗	=	Rare	LC	NS	Rare	LC	261
N. latitarsis (Egger, 1865)	⚥	✗	=	Rare	LC	NS		LC	261
N. pubescens Delucchi & PschornWalcher, 1955	⚥	✗	=	Scarce	LC	NS		LC	260
N. verrucula (Collin, 1931)	⚥	✗	=	Rare	DD			LC	261
N. vitripennis (Meigen, 1822)	⚥	✗	=	Local	LC		Rare	LC	260
Pipiza									
P. austriaca Meigen, 1822	🔍	❓	=	Frequent	LC		Local	LC	256
P. fasciata Meigen, 1822	⚥	✗	▼	Scarce	LC			LC	256
P. festiva Meigen, 1822	⚥	✗					Rare	LC	256
P. lugubris (Fabricius, 1775)	⚥	❓	▼	Local	LC	NS		LC	256
P. luteitarsis Zetterstedt, 1843	🔍	❓	▼	Frequent	LC		Rare	LC	258
P. noctiluca (Linnaeus, 1758)	🔍	✗	▼	Frequent	LC		Local	LC	258
P. notata Meigen, 1822	⚥	✗	▼	Local	LC		Rare	LC	256
Pipizella									
P. maculipennis (Meigen, 1822)	⚥	✗	=	Scarce	LC	NS		LC	262
P. viduata (Meigen, 1822)	⚥	✗	=	Widespread	LC		Local	LC	262
P. virens (Fabricius, 1805)	⚥	✗	=	Frequent	LC			LC	262
Trichopsomyia									
T. flavitarsis (Meigen, 1822)	⚥	❓	▼	Frequent	LC		Frequent	LC	264
T. lucida (Meigen, 1822)	⚥	✗	–	Rare	LC			VU	265
Triglyphus									
T. primus Loew, 1840	⚥	❓	=	Local	LC	NS		LC	264
Tribe: SERICOMYIINI									
Sericomyia									
S. lappona (Linnaeus, 1758)	👁	📷	=	Frequent	LC		Frequent	LC	268
S. silentis (Harris, 1776)	👁	📷	=	Widespread	LC		Widespread	LC	268
S. superbiens (Müller, 1776)	👁	📷	▼	Frequent	LC		Frequent	LC	266

LIST OF BRITISH AND IRISH HOVERFLIES

Name	ID	Photo	TR	Frequency (GB)	GB	BRS	Frequency (Ir)	Eu	Page
Tribe: VOLUCELLINI									
Volucella									
V. bombylans (Linnaeus, 1758)	👁	📷	▲	Widespread	LC		Widespread	LC	270
V. inanis (Linnaeus, 1758)	👁	📷	▲	Widespread	LC			LC	274
V. inflata (Fabricius, 1794)	👁	📷	=	Frequent	LC			LC	272
V. pellucens (Linnaeus, 1758)	👁	📷	=	Widespread	LC		Widespread	LC	272
V. zonaria (Poda, 1761)	👁	📷	▲	Widespread	LC			LC	274
Tribe: XYLOTINI									
Blera									
B. fallax (Linnaeus, 1758)	👁	📷	–	Rare	CR			LC	278
Brachypalpoides									
B. lentus (Meigen, 1822)	👁	📷	▼	Frequent	LC		Scarce	LC	280
Brachypalpus									
B. laphriformis (Fallén, 1816)	🔍	📷	=	Local	LC		Rare	LC	282
Caliprobola									
C. speciosa (Rossi, 1790)	👁	📷	–	Rare	NT			LC	280
Chalcosyrphus									
C. eunotus (Loew, 1873)	🔍	❓	▲	Scarce	LC	NS		VU	284
C. nemorum (Fabricius, 1805)	🔍	❓	=	Frequent	LC		Local	LC	284
C. piger (Fabricius, 1794)	🔍	📷	–	Rare	NE			LC	284
Criorhina									
C. asilica (Fallén, 1816)	🔍	📷	=	Frequent	LC			LC	292
C. berberina (Fabricius, 1805)	🔍	📷	▼	Frequent	LC		Local	LC	294
C. floccosa (Meigen, 1822)	🔍	📷	=	Frequent	LC		Local	LC	294
C. ranunculi (Panzer, 1804)	👁	📷	▲	Frequent	LC		Scarce	LC	292
Pocota									
P. personata (Harris, 1780)	👁	📷	=	Scarce	LC	NS		LC	296
Syritta									
S. pipiens (Linnaeus, 1758)	👁	📷	=	Widespread	LC		Widespread	LC	290
Tropidia									
T. scita (Harris, 1780)	👁	📷	=	Frequent	LC		Frequent	LC	290
Xylota									
X. abiens Meigen, 1822	♂	✕	▼	Scarce	LC	NS	Rare	LC	286
X. florum (Fabricius, 1805)	♂	✕	▼	Local	LC		Scarce	LC	286
X. jakutorum Bagachanova, 1980	🔍	❓	▲	Frequent	LC		Frequent	LC	286
X. segnis (Linnaeus, 1758)	👁	📷	=	Widespread	LC		Widespread	LC	288
X. sylvarum (Linnaeus, 1758)	👁	📷	▼	Widespread	LC		Frequent	LC	288
X. tarda Meigen, 1822	♂	❓	▼	Local	LC	NS	Rare	LC	286
X. xanthocnema Collin, 1939	♂	✕	▼	Local	LC	NS			286

Sub-family: MICRODONTINAE									
Microdon									
M. analis (Macquart, 1842)	🔍	📷	=	Local	LC	NS	Rare	NT	298
M. devius (Linnaeus, 1761)	🔍	❓	=	Scarce	NT			NT	300
M. mutabilis (Linnaeus, 1758)	🔍	✕	–	Local	DD		Rare	VU	300
M. myrmicae Schonrogge et al., 2002	🔍	✕	▲	Local	LC		Rare	VU	300

Photographing hoverflies

Cameras

Digital cameras can be divided into three categories: **Compact cameras**, **Digital Single Lens Reflex (DSLR) cameras** (including **mirrorless cameras**) and **Bridge Cameras**.

Compact cameras are small, point-and-shoot, pocket cameras. They often have a 'macro mode' to tackle small targets and can produce very good images. However, they have disadvantages that make them less suitable for hoverflies. Most have an LCD panel on the back that shows what the lens is 'seeing' instead of a viewfinder. In bright, outdoor conditions it is often difficult to see this image clearly enough to be able to compose and ensure correct focus. They tend to be highly automated and give the user limited control over how the picture is taken. There may be an option to control the area on which the camera's autofocus system works, which is often presented as a rectangle, displayed on the LCD screen, that you can manoeuvre over the area on which you want to focus. If this facility is available, it is usually best to focus on the insect's eyes. Finally, you will generally have to get the camera very close (*i.e.* a few centimetres) to a hoverfly to get an image that fills the frame. This can be a challenge and demands good fieldcraft! Such cameras are increasingly built into mobile phones, which are also capable of producing good images.

Focusing on the head of the hoverfly *Myathropa florea* using the LCD screen of a compact camera.

DSLR cameras are ideal for insect photography, but are expensive, bulky and require knowledge and experience to use effectively. The big advantages are that the viewfinder allows you to see exactly what the lens is seeing, and that the lenses are interchangeable, allowing the use of gear designed for close-ups. DSLRs also give you full control: you can get the exposure and composition you want – providing you know what you are doing! Popular recent variants on the DSLR are **mirrorless cameras**. These are very similar in appearance to DSLRs and share the ability to change lenses and use a wide range of accessories. In a DSLR, an angled mirror is positioned behind the lens and reflects the light into an optical viewfinder. When the button is pressed to take a picture, the mirror flips up out of the way before the shutter is opened. The mirror box is quite bulky and complex. In contrast, in a mirrorless camera, an LCD screen in the viewfinder displays what the camera's sensor is 'seeing', so the mirror box and optical viewfinder can be dispensed with. Consequently, this type of camera tends to be slightly smaller and lighter than an equivalent DSLR. An electronic viewfinder has advantages for insect photography. When close-up gear is used the amount of light entering the camera is reduced, which can lead to a rather dim image in an optical viewfinder, making it difficult to frame and focus the picture. However, in a mirrorless camera, the electronics maintain a bright viewfinder image. This is a substantial benefit when trying to photograph small subjects where high magnification is necessary.

Bridge cameras are intermediate between a compact camera and a DSLR. They usually have an electronic viewfinder like a mirrorless camera, which works well in bright daylight. They usually have a fixed lens, but this is often a 'super-zoom' that offers a very wide range of focal lengths with good macro capabilities. Physically, they are also intermediate: not as small as a compact, but not as heavy or bulky as a DSLR.

Magnification

The Drone Fly *Eristalis tenax* pictured here was about 15 mm long in life. As printed here it is 38 mm long (*i.e.* magnified about 2½×). But it could have been printed so that it filled the whole page – in which case it would appear about 110 mm long (*i.e.* magnified 7×). It is easy to see that 'magnification' depends upon the way in which the image is viewed and is, in consequence, not a very useful measure. A much more useful way to think about it is the size of the image on the camera's sensor.

A 'macro photo' is one in which the image of the subject is at least life-size on the camera's sensor. The term 'reproduction ratio' is used to describe the scale of the image – a ratio of 1:1 meaning life-size – and is a much more useful way to think about magnification. Successful hoverfly photographs are generally taken at reproduction ratios between half life-size (1:2) and two or three times life-size (2:1 or 3:1).

The image of the APS-C sensor, as used in some DSLRs, shows how a Drone Fly *Eristalis tenax* would appear at life-size (reproduction ratio 1:1)

There are several ways of achieving magnification:

Supplementary close-up lenses are screwed into the filter mount at the front of the lens. The magnifying power of such lenses is measured in dioptres. If your camera has a fixed lens, then this is really the only possibility. Simple cameras may lack a screw thread at the front of the lens, but a push fit, or even some adhesive tape, can always be contrived. The disadvantage of close-up lenses is that the extra layers of glass in front of the lens mean some degradation in the quality of the image is likely – especially if using cheaper ones.

When you focus a camera on a close object, the front of the lens moves farther out as you turn the focusing ring. How close you can focus is determined by how far out it can move. If the camera has interchangeable lenses, then the lens can be moved further away from the camera by inserting extension tubes. Modern cameras have autofocus and exposure systems that require electrical connections between the camera and lens. Extension tubes must maintain these connections. Optics dictate that, for a lens focused at infinity, an extension equal to the focal length is needed to produce an image with a reproduction ratio of 1:1. Since the 'standard lens' supplied with many DSLRs usually has a focal length of around 50 mm, extension tubes tend to come in sets of about that length, typically supplied as three tubes of different sizes so that a number of

Extension tubes

combinations are possible. Extension tubes are just empty tubes, and therefore have no effect on image quality. However, moving the lens further from the sensor spreads the light coming through it over a larger area, so the image is dimmer; extension tubes to get to 1:1 will dim the image by two stops (see *below* for an explanation of what this means).

Macro lenses are specifically designed to perform best when the subject is extremely close. They are very convenient and produce superb results, but they are expensive. It may be advisable to buy a moderately priced set of extension tubes first and experiment with those before advancing to a macro lens.

Camera shake. This whole shot of *Baccha elongata* is blurred and the highlights extend as streaks.

Exposure

Cameras have three controls that affect the exposure:

The **shutter speed** is the length of time for which light is allowed into the camera, measured as a fraction of a second (*e.g.* 1/16, 1/250, *etc.*). The longer the shutter is open, the more light falls on the camera's sensor. Since hoverflies are active and fast-moving animals, a short exposure to freeze motion is an advantage. Also, the camera must be kept still during the exposure to avoid 'camera shake' – where the whole picture is blurred because the camera moved during the exposure. A fast exposure is desirable in order to stop the movement of both the subject and the camera.

Insufficient depth of field. The thorax of this *Hammerschmidtia* is in focus, but the head is not because the DoF is too small. Pictures of animals are rarely acceptable if the eyes are not sharp!

The **aperture** is the size of the hole through which light enters the camera. It is measured in f-stops (*e.g.* F6·3, F16, *etc.*). (Note: The smaller the hole, the bigger the f-stop number – so F16 denotes a smaller aperture than F8.) A larger aperture means a brighter image on the sensor and hence a stronger signal. However, the size of the aperture also affects

Photo of *Syrphus ribesii* drinking taken at high ISO (1,600). The magnified insert shows the noise visible in the dark areas.

the distance over which focus can be attained (termed the **depth of field** – DoF) – which increases as the size of the aperture decreases (and as the f-stop number increases). This is important when photographing insects for identification purposes, where the picture needs to be in focus.

Finally, the **sensitivity** of the camera's image sensor can be changed. This is measured by an ISO number. Lower ISO numbers (*e.g.* ISO 100) indicate less sensitivity – more light is needed to form a usable image. As the ISO number is raised (say to ISO 400 or 800), the amplification in the imaging system's electronics increases and an image can be recorded with less light. Some modern DSLRs will allow very high ISO settings (up to ISO 102,400). So why not always use the highest ISO available? The problem is image 'noise'. This appears as a random scatter of lighter or darker pixels in the image which are particularly obvious in dark areas. If you turn up the ISO too far, the picture will be degraded by noise. A low ISO settings is desirable for noise-free images.

When you photograph a hoverfly, there is only so much light available. The ideal is a fast shutter speed to stop motion, a small aperture to give as much DoF as possible and a low ISO setting to avoid noise. There is rarely enough light to achieve all of these (otherwise it would be too easy!), so compromise is needed.

Each of the scales of measurements is calibrated in halving or doubling of the amount of light. A shutter speed of 1/125 lets in about half as much light as 1/60. Similarly, changing the f-stop by one increment halves or doubles the area of the aperture – F8 lets in twice as much light as F11. Equally, you need twice as much light to form an image at ISO 100 as you would at ISO 200. For example, if an exposure of 1/125 at F8 using ISO 200 gives a correctly exposed image, the same exposure would be obtained by 1/60 at F11 (shutter open for twice as long, but the aperture half the size) or keep 1/125 shutter speed and Increase the ISO to 400 (half the size of the aperture, but double the sensitivity of the sensor). One of a photographer's skills is juggling these settings to get the desired result.

A point-and-shoot camera (or a DSLR in 'Auto' mode) will make these decisions for you. Modern cameras can make a pretty good job of taking 'typical' pictures but may not cope so well with close-ups. Point-and-shoot cameras usually offer a range of picture styles, 'close-up' (often denoted by a flower icon) usually being one of them. This setting helps the metering system and increases the chance of a good shot. Use it for hoverfly pictures if your camera offers it.

More sophisticated cameras allow you to take control. Hoverfly pictures on a DSLR will most likely be taken using 'Aperture Priority' mode. In this mode, you set the aperture (to achieve the necessary DoF) and the camera's meter will suggest a shutter speed appropriate for the available light. You must decide whether that shutter speed is fast enough and, if not, increase the ISO to increase it.

Electronic flash

Electronic flash provides a way out of the dilemma in which a fast shutter speed is needed to stop motion, but a small aperture is required to achieve sufficient DoF. Flash guns provide a bright source of light that can be positioned close to the subject, thereby allowing a small aperture to be used, giving a greater DoF. The way in which an electronic flash gun works ensures that the flash of light it produces is very brief indeed – typically 1/1,000th of a second or less. This is short enough to counter camera shake and stop most motion. (But a hovering hoverfly typically beats its wings around 300 times per second. In 1/1,000th of a second it will go through one-third of a wing-beat and the wings will still show motion blur!)

Electronic flash can, however, result in a brightly lit subject against a dark, or even completely black background because the flash does not reach far enough to illuminate the background. This is not a problem when photographing a hoverfly resting on a flower or leaf that fills the

Two pictures of male *Helophilus pendulus* taken in the same location with a 100mm macro lens at around 1:1 and at F11. The picture on the left was taken in bright daylight (1/160, 400 ISO) – note the strong shadows. The picture on the right was taken using the twin flash set-up shown *below* (1/250, 100 ISO) – note how the flashes have filled in for each other resulting in little shadowing whilst the colours are stronger and more contrasted. The background (a yellow polythene bowl) completely fills the frame in the right-hand shot, so there is no problem with a black background.

whole field of view. The harshness of the lighting can also be ameliorated by fitting the flashes with suitable diffusers.

Another use for electronic flash is to fill in shadows in bright sunshine. A hoverfly sitting in full sun will have a strong, dark shadow. A bit of flash to soften the shadows can make for a much better picture.

Support

Another way of avoiding camera shake is to support the camera using a tripod, monopod or beanbag. Longer focal length macro lenses and close-up flash can get rather heavy, so extra support is all the more necessary.

Macro lens and close-up flash set-up.

Tripods are awkward to use when stalking hoverflies but can be erected where you expect flies to appear (*e.g.* next to a patch of flowers). A monopod is much easier to use than a tripod when stalking and provides support while still enabling you to focus by rocking backwards and forwards. Another option is to trigger the camera using a remote release when the subject comes into the pre-set field of view.

A beanbag placed on the ground can support the camera when you need to get very low down. In this situation, it can be rather difficult to get your head in position to see through the viewfinder – but this can be overcome with a right-angled attachment fitted on the viewfinder. Some cameras have an articulated LCD screen on the back, which is an advantage when composing and focusing in these circumstances.

Fieldcraft

An insect the size of a hoverfly usually requires around a life-size (1:1) image to fill the camera's frame – a little less for large species like *Volucella*, but perhaps twice or even three times life-size for small species like *Melanostoma*. This requires a close approach, although how close depends upon the focal length of the lens.

Getting close to a hoverfly in the field can be a challenge but the following tips may help:

- Approach slowly and smoothly, avoiding sudden movements – hoverflies are very sensitive to movement.
- Do not let your shadow fall over the fly – that is a sure way of scaring it off!
- Be very careful not to disturb the vegetation on which the fly has settled.
- If a hoverfly is working its way over a patch of flowers or a large umbel, then position yourself at the nearest convenient spot and wait for the fly to come to you. Males of many species will return repeatedly to a perch when they are searching for females. Position yourself close to one of these perches and be patient.
- Hoverflies rarely show much reaction to flash but may respond to the infra-red beam sometimes used by compact and bridge cameras to assist autofocus.
- Turn autofocus off and focus by rocking gently back and forth. Autofocus systems are often fooled by stems and the like nearer and farther from the subject, leading to 'seeking', where the focus mechanism keeps shifting from one point to another, which can lose you the shot!
- Do not try to bring the insect into focus and then freeze. Instead, try to move smoothly through the point of sharpest focus, pressing the shutter at the correct moment without pausing. If your camera can take multiple shots, take a short burst of, say, three. This both increases the chance that one will be taken at the optimum DoF, but also means that the slight jerk as the shutter button is pressed only affects the first shot.
- It is usually best to get the animal's eyes in focus, as you can afford to have wing-tips, the ends of the legs, etc. going out of focus without spoiling the picture. Remember that about half the DoF is in front of the point of sharpest focus and half behind. Choose the point where you focus with this in mind.

Finally, it is important to think about the vegetation you may be trampling in order to get into position for a photograph. If there is any risk of damaging the vegetation, especially on protected sites (*e.g.* chalk grassland, tall fen, *etc.*) then stop – the welfare of the wildlife and habitats must come first!

Further reading

Read the manual and make sure you know how your camera works and can put it into close-up mode or adjust the ISO settings, *etc.*

Insect Photography: Art and Techniques by John Bebbington (Crowood Press, 2012) provides a good introduction, whilst *Extreme Close-Up Photography and Focus Stacking* by Julian Cremona (Crowood Press, 2014) covers much higher magnification techniques.

Two internet sites that seem to be well moderated and feature articles by knowledgeable practitioners are Cambridge in Colour (**www.cambridgeincolour.com**) and The Luminous Landscape (**www.luminous-landscape.com**).

Collecting hoverflies

Catching hoverflies

The key piece of equipment required for catching hoverflies is an insect net. These are available from a variety of entomological suppliers and are based on four-fold landing net frames sold to anglers. The bag is made out of a light, rip-stop material. Nets made for catching butterflies and moths tend to be black, but these are not suitable for flies as you cannot see them in the net! Dipterists mostly use white nets. The bag needs to be long enough so that the end folds over the edge of the frame, trapping the fly in a closed pocket.

Using a hand lens to examine a hoverfly.

These frames have a standard-sized screw fitting and a variety of poles can be used interchangeably. They are typically sold with an aluminium pole about 1 m long. These are fine for most purposes, but a longer handle can be useful, especially early in the year, when trying to catch hoverflies visiting blossom such as willows and Blackthorn. Some species, such as *Cheilosia grossa*, *Criorhina ranunculi* and *Platycheirus discimanus*, tend to stay high up and out of reach – a telescopic landing net handle from a fishing tackle shop can extend to 3 m or more and allow you to reach the tops of bushes.

Hoverflies have very good reactions and can move extremely fast, so catching them can be a challenge. A stroke of the net that comes from behind and underneath is usually the most effective technique. Small species such as *Melanostoma* and *Platycheirus* are hard to spot and are often best found by 'sweeping'. This involves sweeping the net gently back and forth through grass and other low vegetation. Watch out for brambles, briars, other thorny shrubs and barbed wire to avoid inadvertently shredding your net!

Individual hoverflies can be picked out between the thumb and fingers for close examination without harming them. However, if you want to keep the hoverfly for later examination, you need to transfer it to a container. Small, transparent specimen tubes (glass or polystyrene) with polythene caps are useful for this. Take the top off the tube and place it over the hoverfly in the net so that it is trapped in the tube against the netting. Then shake the fly into the tube and slide the cap on. This becomes fairly routine with practice, but it is inevitably the best catch that manages to escape!

The pooter is a better option for extracting smaller hoverflies, especially after you have been sweeping. There are several designs, but the principle is the same. There is a closed, transparent container with two tubes leading into it through a rubber bung. You suck through one and the fly is sucked into the container through the other. The tube you suck through has some gauze over the inner end so that you don't get a mouthful of flies!

A good way to empty your net after sweeping is to first carefully let out the more aggressive and larger catch, like bumblebees, and then to put the net over your head with the end of the bag held so that it is pointing towards the brightest part of the sky. Insects will tend to walk up the netting towards the light, so this keeps them away from

COLLECTING HOVERFLIES

Using a long-handled net.

Using a pooter to extract flies from a standard net.

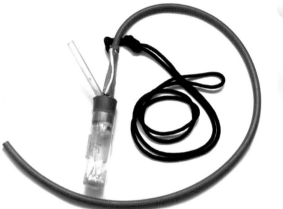

A pooter. Suck through the rubber tube and flies are sucked into the glass specimen bottle through the clear polythene inlet tube. The tube you suck through has gauze over it to avoid flies getting sucked into your mouth. The black cord is a lanyard to avoid losing or dropping the pooter.

Safety considerations:

- Do not poot insects collected off carrion, dung or fungi. You might breathe in spores or unpleasant bacteria! (This is not usually an issue when catching hoverflies.)
- Make sure your pooter tube is cleaned and ventilated before re-use to avoid getting a mouthful of ethyl acetate fumes next time you use it.

the entrance and escape, and also away from your face. You then pick off the hoverflies by sucking them into the pooter.

Hoverflies can be killed by putting a couple of drops of ethyl acetate on a twist of tissue paper and dropping this into the container with them; the vapour knocks them out very quickly. It is advisable to use glass or polythene tubes for this purpose and to avoid putting ethyl acetate into polystyrene tubes as it will dissolve the plastic. Ethyl acetate can be obtained from entomological suppliers, but cannot be sent by post. You will therefore either need to buy it in person at an entomological exhibition or have it sent by a courier service (which makes it rather expensive). An alternative is to put the tubes containing the hoverflies in the freezer for a couple of hours.

Curating specimens

Specimens of hoverflies are best kept dry and pinned using headless, stainless steel micro-pins. These are available from entomological suppliers and come in various thicknesses (designated by letters, where A is the thinnest) and lengths (designated by numbers: 1 = 10 mm, 2 = 15 mm, *etc.*). A1 pins are the shortest and finest and are suitable for pinning small hoverflies like *Platycheirus*; B2 pins are suitable for medium-sized species like *Syrphus*; and C3 pins are suitable for big hoverflies like *Volucella*. Small hoverflies are often pinned on their sides with the pin passing through the thorax. Ideally, the pin should be inserted slightly off the vertical so that it comes out in a slightly different place on the other side of the thorax. This means that any feature it penetrates (and potentially destroys) will survive on at least one side. Bigger hoverflies are often pinned flat with the pin passing from the top to the bottom of the thorax.

Hoverflies need to be pinned when fresh. If you allow your specimen to dry out it will become brittle and pinning it in this state risks parts falling off. Once the micro-pin has been inserted through the hoverfly, spread out its wings and legs so that everything can be seen and hold the various parts in place with other pins while the specimen dries out. This is also the stage at which to tease out a male's genitalia so it is visible. Again, the genital capsule can be hinged out and held in place with a pin until it dries and will then

A male specimen of the small hoverfly *Sphaerophoria* that has been side-pinned using an A1 micro-pin and then staged and mounted on a continental pin with data and determination labels.

stay in the position required. Once the specimen is dry, the temporary positioning pins can be removed.

For long-term storage, the micro-pinned specimen should be pinned to a 'stage'; this is a strip of vibration-absorbing material such as plastazote or polyporus (the Birch Bracket fungus), which is mounted on a much longer pin. Continental length stainless steel pins from entomological suppliers are best for this purpose. The specimen should always be handled by the large pin to avoid damage and never directly touched or handled by the micro-pin.

The most important part of any specimen is the labels. These are also kept on the large pin and record the details about the specimen. There will normally be (at least) two labels: a data label, including where and when the specimen was caught; and a 'determination' label which gives the identification and the name of the person who identified it.

Why keep a collection?

There are a number of good reasons for keeping specimens of hoverflies in a collection.

Vouchers: You can prove to yourself and others that you got the identification correct. This is especially important when you are starting off. The best way to learn is to try naming some specimens and then to get them checked by an expert. Nevertheless, however expert you may become, it remains important to keep vouchers for difficult and uncertain identifications.

Comparison: The identification process often involves comparative judgements (*e.g.* bigger than …; eye hairs darker than …). In these cases, it is very useful to have material you have previously identified to hand so that you can remind yourself what the various options look like.

Species splits: Knowledge of hoverflies is growing all the time and there have been a number of recent taxonomic revisions which have led to species being 'split'. There are also further such revisions under consideration. When a species is split into two or more new species, it is only possible to confirm your previous identifications if you have kept specimens that can be re-examined.

One of the authors, Roger Morris, dealing with the day's catch during a Dipterists Forum summer field meeting.

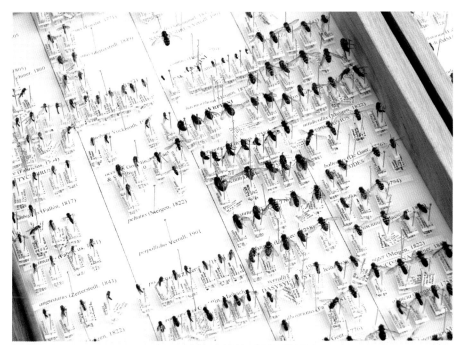

A collection of voucher specimens in a plastazote-lined wooden store box.

The ethics of collecting

Many naturalists frown upon the collection of specimens, the prevailing ethos being 'leave nothing but footprints, take nothing but photographs'. As discussed previously, however, this approach does not work for a large proportion of Diptera, or indeed for many other invertebrates, as it is often necessary to examine a dead specimen rather closely in order to ensure an accurate identification.

So is there a potential conflict between collection and conservation? The authors of this book have both spent most of their working lives employed by conservation organisations and do not consider there to be a conflict. Whilst we recognise that a few rare species have such restricted breeding sites that collecting could be damaging, we believe that insect faunas are at far greater risk through not knowing which sites are important and how they should be managed. The collection of a few voucher specimens is extremely unlikely to be damaging to hoverfly populations and, in our view, responsible collecting that adds to our knowledge of the distribution and biology of hoverflies should be encouraged. However, we would stress that if an animal has died to generate an identification then use should be made of the resulting record! It should be lodged with the Recording Scheme and subsequently made available to conservation organisations and researchers. Although not a good reason to condone collecting, it is worth bearing in mind that you will kill many more hoverflies on the front of your car driving to and from a site than you are ever likely to collect.

The Joint Committee for the Conservation of British Invertebrates (JCCBI) first published A Code of Conduct for Collecting Insects and Other Invertebrates in 1969 and this was most recently revised in 2002. It is available from www.amentsoc.org/publications/online/collecting-code.html and anyone wishing to collect hoverflies should familiarise themselves with this Code and adhere to its principles.

Legislation and conservation

Only a few insect species are afforded some degree of legal protection in Britain, the principal legal instrument being the Wildlife and Countryside Act (1981). As a signatory to the Rio Convention on Biological Diversity in 1992, the UK government published its first Biodiversity Action Plan (UKBAP) in 1994. Species were selected for conservation action on the basis of conservation decline, highly localised distribution, international threat and importance, and Red Data Book status. The UK Biodiversity Action Plan currently lists seven hoverfly species as priorities for conservation action: *Blera fallax*, *Callicera spinolae*, *Chrysotoxum octomaculatum*, *Doros profuges*, *Eristalis cryptarum*, *Hammerschmidtia ferruginea* and *Myolepta potens*.

Single-species studies have now been conducted on all seven UKBAP Priority Species, with varying degrees of success. Most success has been achieved with the Pine Hoverfly *Blera fallax* and the Aspen Hoverfly *Hammerschmidtia ferruginea*, the ecology of both of which is now well known as a consequence of work by the Malloch Society and, latterly, through a PhD study undertaken by Ellen Rotheray. Both *B. fallax* and *H. ferruginea* have been shown to have extremely small populations and to be at considerable risk of extinction. However, both have been shown to respond well to active conservation management, with new breeding sites created by judicious use of the chainsaw to create new rot-holes and fallen timber respectively. *H. ferruginea* is a flagship species for a wider assemblage of insects associated with Aspen woods in Scotland. These woods are mostly small and many are heavily grazed, showing few signs of recruitment of new trees. They are potentially vulnerable to other conservation initiatives such as the reintroduction of European Beaver.

Useful population studies of *Eristalis cryptarum* have shown this species to be confined to a very small part of southern Dartmoor, but the larvae have still not been found despite careful work over several years. The larvae of both *Myolepta potens* and *Callicera spinolae* are known to occur in water-filled rot-holes, but these species are still highly elusive and relatively little is known about their distribution or the factors that make them so restricted. *Doros profuges* and *Chrysotoxum octomaculatum* have proved to be particularly elusive and little new information has emerged.

At present, our understanding of the status of hoverflies in Ireland is rudimentary. As mentioned previously, there are currently insufficient data to be able to undertake an analysis of trends, which is one of the key elements in assigning a threat status to a species according to IUCN guidelines. Although efforts are being made to develop Irish Red Lists for insect groups such as hoverflies, at the time of writing this had not been published.

Eristalis cryptarum

Recording hoverflies

The Hoverfly Recording Scheme (HRS) is one of the more active wildlife recording schemes in Britain. It covers the British mainland and islands, together with the Isle of Man. It does not cover Northern Ireland, the Republic of Ireland or British Protectorates such as the Channel Islands. Recorders in the Republic of Ireland and Northern Ireland should send records to the National Biodiversity Data Centre in County Waterford and CEDaR, respectively (see *page 338*). As far as we know, there is no formal recording scheme in the Channel Islands, which perhaps is an opportunity for somebody to build interest in hoverflies there.

Data for recording schemes such as the HRS are welcomed in computer-readable or paper formats. Recorders who use packages such as MapMate and Recorder can be accommodated readily; otherwise, data are preferred in spreadsheet format that include the basic headings shown *below*: (red = required; green = optional). As increasing numbers of people take an interest in larvae, it is important to ensure that larval records are carefully flagged with the date on which the larva/pupa is found entered in the 'Date' column, the life stage logged in the 'Sex/life stage' column and the date the adult emerged in the 'Notes & comments' column.

Species name	Location name	OS grid reference	Date	Recorder's name	Sex/life stage	Number	Notes & comments

In addition, records posted on the UK Hoverflies Facebook page are extracted if they are not to be submitted by other methods such as iRecord, iNaturalist, SyrphBoard, *etc.* (for website addresses see *page 338*).

Data compiled by recording schemes are used in a growing number of ways that greatly extend their traditional use in species distribution mapping (see *Putting data to good use* on *pages 324–332*). One of the most important uses is in the compilation of Red Lists (see *page 302*). HRS data have been used to compile Britain's Red List (Ball & Morris, 2014) and were also supplied to several of the authors of the European Red List (Vujić *et al.*, 2022) (see *Further reading* on *pages 337–338*).

Modern analyses that generate trends and model potential occupancy are potentially influenced by the degree to which the data cover both the breadth of species occurrence and their relative frequency within the same year. Highly selective recording such as only noting 'interesting' species or the first/last date for a species in a year introduce serious biases into the dataset, so it is important to record all the hoverflies seen, even if this means a sequence of daily records for a 'common' species from the same locality.

Records of even the commonest species can be important because more of them are received and therefore it is more likely that it will be possible to detect changes. Long-term trend analysis also depends upon the robustness of the dataset; consequently, the Recording Scheme organisers scrutinise data submitted and may seek confirmation of the identity of trickier species. Records of species that are considered to be difficult to identify cannot be accepted without supporting evidence (*e.g.* a photograph and/or a preserved specimen). An identification service is offered for specimens that require microscopic examination. On occasion, large batches of specimens can be identified if required for research projects.

Putting data to good use

Since the first edition of this book was published in 2013, the use of biological records by scientists to evaluate the impact of issues such as the decline of pollinators and climate change has become more prevalent. Universities and other academic institutions rely upon sources such as the Hoverfly Recording Scheme (HRS) to provide long-term data of this type. Good examples of recent outputs include the triennial *State of Nature* reports produced by a collaboration of statutory and non-governmental organisations, species status reviews published by the conservation agencies, and a host of papers on pollinator and other wildlife trends published by the UK Centre for Ecology & Hydrology and several universities. We, too, have produced a range of analyses that have appeared in various journals. Most of the more relevant recent ones are listed in the section on *Further reading (page 337)*.

Trends

There have been recent reports covering a variety of insect groups which have shown that many species are in decline and that the overall abundance of insects is also decreasing. British hoverflies are no exception, and it appears that the majority of species show evidence of decline and that the overall frequency with which hoverflies are recorded relative to the amount of recording effort has also declined. We have not been able to collect data about the abundance of hoverflies in any systematic way, so we cannot offer any analysis of abundance changes, but our field collecting experience does suggest a decline here also.

These changes, however, are not uniform. Figure 2 shows the result of an analysis in which the data are split into five regions (SE England, SW England & S Wales, NE England, NW England & N Wales, and Scotland) based on their location on the Ordnance Survey (OS) National Grid.

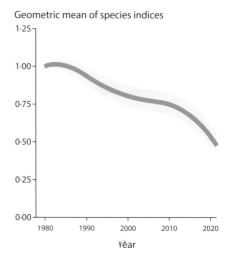

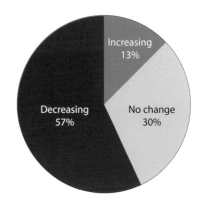

Figure 1: Trends in British Hoverflies, 1980–2021, calculated using Frescalo (Hill, 2012).
The graph (*top*) shows the relative frequency averaged across all species. The pie-chart (*above*) shows the proportion of species in which the relative frequency has declined significantly, increased significantly or for which there is no significant change. There were 228 species for which it was judged that there were sufficient data to contribute to these plots.

The largest proportion of species are in decline in south-east England and the least in Scotland. The reasons for these differences are not fully understood, but we believe that higher rates of decline in eastern England are correlated with the greater concentration of intensified agriculture and urban development, exacerbated by climate change resulting in extreme events such as droughts that have impacted this part of the country most severely.

Unfortunately, at present there are insufficient data for Ireland to enable the calculation of trends using similar techniques.

Trends are also not distributed uniformly across habitat. Figure 3 shows trends categorised by the habitat preferences of species. These results suggest that wetland-associated species show the greatest proportion of declines, with those associated with coniferous woodland showing a very similar proportion. Broadleaved woodland species show the lowest proportion of declines (but, at the same time, no increasing species). Again, the declines in wetland species are consistent with extreme weather such as long, dry summers having an impact.

Changes in recorder behaviour

We have noticed a long-term trend for recorders to concentrate on commoner and more obvious species, with those that are more difficult to find and identify, especially when identification requires microscopic examination, becoming less well recorded (Figure 4). Since 2013, this tendency has accelerated greatly and can be related to the establishment of the UK Hoverflies Facebook page (see *page 338*). This interactive facility has encouraged many more recorders to become involved and has led to the majority of records now coming from photos that are posted online. The numbers of records received each year (Figure 5) and the number of people who have submitted records (Figure 6) has increased massively.

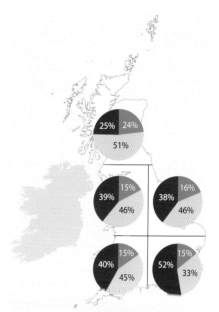

Figure 2: Regional trends.
Proportion of species showing a significant decline, increase or no significant change in five regions of Great Britain, 1980–2021. The number of species judged to have sufficient data to contribute to the analysis were SE 205; SW 121; NE 171; NE 150; Scotland 145.

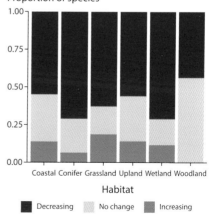

Figure 3: Trends by habitat.
The proportion of species showing a significant decline, increase or no change in the habitat preference of the species.

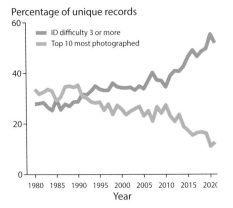

Figure 4: Proportion of records received per year of the ten most-photographed species, compared with the proportion for species that are judged to be difficult to identify (☝).

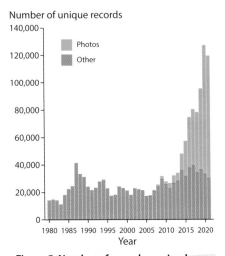

Figure 5: Number of records received per year, showing how many were based on photos submitted online.

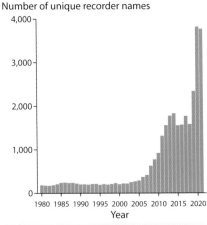

Figure 6: Number of different recorders from whom records were received each year.

These photos are mostly of the larger, more obvious and photogenic species.

This change in the behaviour of recorders has impacted our ability to detect trends. Figure 7 compares trends for two species: *Eristalis pertinax* (p. 224) is large, obvious and our second most-frequently photographed species; *Platycheirus clypeatus* (p. 90) is small, black and not obviously a hoverfly to inexperienced recorders. It is found most frequently by sweeping damp vegetation and is not very often photographed (it is our 133rd most-photographed hoverfly). *E. pertinax* shows an increasing trend but, because it appears so often in photos posted online, its relative frequency and hence its increasing trend is exaggerated. Conversely, *P. clypeatus* is usually overlooked and few photographs are posted online. This leads to its relative frequency being reduced, thus exaggerating the decline. For this reason, records based on photos were excluded in preparing Figure 1, Figure 2 and Figure 3.

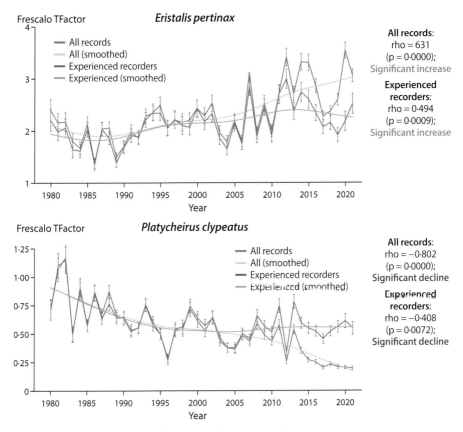

Figure 7: Comparison of trends in two species: *Eristalis pertinax* and *Platycheirus clypeatus*.
E. pertinax is one of the most frequently photographed species. *P. clypeatus* is small, black and has few photos submitted – most often recorded by sweeping in damp places. The plots compare an analysis based on 'all records' (blue) and on a subset of records from 'experienced recorders' (those who have submitted records of at least 100 different species). The large increase in the proportion of records from photos posted online has tended to increase the relative frequency of common and obvious species whilst decreasing that of the obscure and less photographed species.

♂ *Eristalis pertinax*

♂ *Platycheirus clypeatus*

♀ *Volucella zonaria* ♀ *Cheilosia caerulescens*

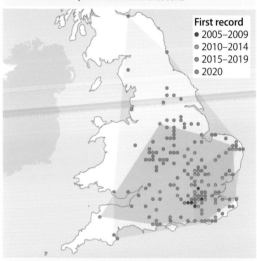

First record
- 2005–2009
- 2010–2014
- 2015–2019
- 2020

Figure 9: Spread of *Cheilosia caerulescens* through Great Britain since its introduction in 2006.
The shaded areas show the limits of the distribution in five-year periods.

Range change

The ranges of some species are expanding rapidly. The spectacular hornet mimics *Volucella zonaria* and *V. inanis* (p. 274) are good examples. Both were restricted to the extreme south of England, especially the London basin, before 1995 and have expanded rapidly northwards in the last 25 years. Figure 8 illustrates the expansion of *V. zonaria* from 1980 to 1995 and the most recent five-year period, 2018–2022. It seems likely that the warming climate has allowed these two species to expand their range: until around 1995 they were largely limited to the 'urban heat island' of the London area but, since then, increasing temperatures have allowed them to spread.

Figure 8: Expansion in the range of *Volucella zonaria*.
The hectads from which records were received in 1980–1995 and 2018–2022. These two periods were chosen because the total number of records submitted to HRS were very similar.

Cheilosia caerulescens (p. 176) (Figure 9) is a rather different case: it originally occurred in high mountains, such as the Alps, in western Europe where its larvae mine the leaves of houseleeks. Over about the last 50 years it has been spreading westwards using plants in garden rockeries, on walls and roofs, *etc*. By 2000 it was widespread in the Netherlands and Belgium, especially in urban areas, and was anticipated to reach the UK via imported plants sold through the horticultural trade. It was first recorded in England in 2006 when it was noticed at two localities in Surrey, mining the leaves of houseleeks that had been purchased from garden centres. These were probably not actually the first to occur, but the plants happened to have been purchased by someone interested in hoverflies! Since then, the species has spread rapidly throughout England and reached Scotland and south Wales by 2021.

The situation with *Rhingia rostrata* (p. 194) (Figure 10) is a much more mysterious case. This species was considered a real rarity before 2000, with the few records coming mainly from the Weald of Sussex and from the Forest of Dean westwards through south Wales. In the first British Insect Red Data Book (Shirt, 1987) it was rated 'RDB2 – Vulnerable' and in the 1991 status review (Falk, 1991) it was categorised as 'RDB3 – Rare'. In the early 2000s, however, it began to be recorded more frequently, especially in the Forest of Dean area and northwards along the Welsh borders. In about 2005 it was found in the East Midlands for the first time ever. Subsequently, it has spread eastwards and northwards throughout England and Wales and is now as frequent as its common relative, *R. campestris* (p. 194), in many areas. Similar range expansions have also been observed in other parts of north-west Europe and we have no idea why!

Where species are declining, there is often little discernible effect on their range for quite a long period. Such species become less frequent throughout their range as local colonies disappear, but there is a delay of years or decades before this is evident from 10km square 'dotmaps'. Figure 11 illustrates this for *Platycheirus fulviventris* (see *p. 92*), a species that has shown a strong and sustained decline over the last 40 years. Whilst the range has not changed, it has become more sparsely distributed.

♀ *Rhingia rostrata*

pre-1990

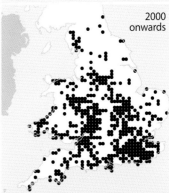

2000 onwards

Figure 10: Expansion in the range of *Rhingia rostrata*.
The top plot shows records pre-1990 and the bottom plot shows records from 2000 onwards.

PUTTING DATA TO GOOD USE

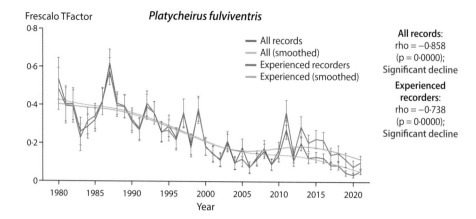

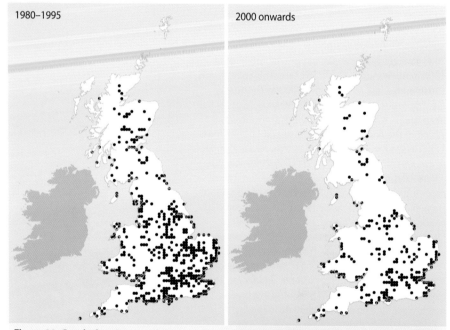

Figure 11: Graph showing trends in recording of *Platycheirus fulviventris* and maps showing distribution of records uring the periods 1980–1990 (LEFT) and 2020 onwards (RIGHT). These two periods were chosen because the total number of records submitted to the HRS during each of them was very similar.

♂ *Platycheirus fulviventris*

Leucozona glaucia (p. 118) does not show a long-term decline, but its range does seem to have shifted in a north-westward direction. The distribution map in Figure 12 shows the most recent record in each hectad (blue for old, red-orange for the latest records). This map shows an area of blue dots in south-east England. This is surprising because this has consistently been the most heavily recorded area. On the other hand, most records in the north-west are recent. The right-hand map in Figure 12 shows a comparison of Frescalo-rescaled frequency between two periods: 1980–1994 and 2018–2021. The rescaled frequency attempts to correct for recording effort on the relative frequency with which the species was observed. The difference between these two periods suggests that the relative frequency has declined in the south and east, whilst it has increased in the far north and west and also in the uplands of Wales and northern England. We suspect that there are more species beginning to show this sort of north-westward shift, but, as yet, the evidence is not there.

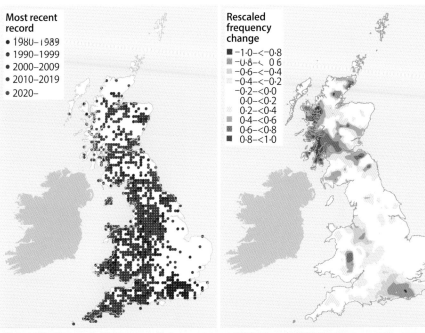

Figure 12: Change in the range of *Leucozona glaucia* between two periods: 1980–1990 and 2018–2021.
LEFT Actual observations, with colours indicating the most recent year for which a record was received for each hectad;
RIGHT The difference in Frescalo-rescaled frequency between these two periods.

♀ *Leucozona glaucia*

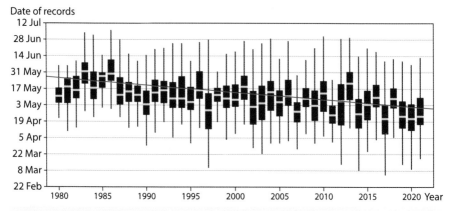

Figure 13: Flight-period of *Epistrophe eligans* from 1980 to 2021.
In this 'Box and whisker plot', the boxes show the date range in which 25–75% of the records received that year fall, with the median date marked by the yellow line. The 'whiskers' (vertical grey lines) show the date range in which 95% of records fall. The red line shows the trend for the median date.

Phenology

Phenology is the study of periodic events in biological life-cycles. There has been considerable interest in the effects of climate change on the timing of events such as the flowering and fruiting times of plants, emergence times of insects, arrival dates of migrant birds, *etc*. We see this among hoverflies too. Figure 13 shows the flight-period of the abundant spring species *Epistrophe eligans* (p. 136). Whilst the flight-period varies from year to year depending on the weather, there has been a clear tendency for it to become earlier and, over the period from 1980 to 2021, it has advanced by nearly a month. It has become not uncommon in recent years for the first records of this species to occur in February!

The emergence time of *E. eligans* seems to be related to temperature – so, not surprisingly, there is also a regional effect with the flight-period being later farther north. Figure 14 compares the flight-period of this species from 2010 to 2021 between the five regions shown in Figure 2 (p. 325).

The emergence of adult flies is often dependent upon the temperature at which their larvae and pupae develop. As a result, flight-periods of the adults will clearly vary according to latitude, longitude and altitude.

♂ *Epistrophe eligans*

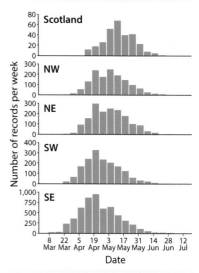

Figure 14: Flight-periods for *Epistrophe eligans* by region, 2010–2021.

Research opportunities

There has been an explosion of interest in 'pollinators' in the past ten years. Whilst laudable, this interest potentially misses the real opportunities to investigate 'pollinator' decline. Although it is true that many hoverflies visit flowers, and many females require pollen to provide the protein needed to facilitate egg maturation, the factors that affect hoverfly abundance are far more complicated. Importantly, it should be remembered that the adult hoverfly is simply the reproductive and dispersive stage.

Larvae, in particular, have the broadest and arguably most important roles in ecosystem regulation. Despite this, surprisingly little is known about the larval stages of hoverflies and about a third of the British and Irish fauna is completely unknown. This stage in the life-cycle is potentially very vulnerable to environmental drivers such as habitat loss, changing temperatures, intensification of soil moisture deficits and declines in humidity during the summer months. Larval stages of many insects are highly susceptible to desiccation and also to drowning – so, heatwaves, droughts and extreme flooding are all potentially highly detrimental at a local or regional level. Moreover, drought in Scotland almost certainly impacts far more quickly where hard rocks and thin soils limit the sponge-like properties of deeper soils.

During a heatwave, it is very noticeable that hoverflies and many other insects largely disappear during the heat of the day. What happens to them? Do they hide away and emerge during cooler conditions, or are they affected by heat stroke and water stress? The limited data that we have suggest that heatwaves lead to a loss of insect activity beyond the immediate period of extreme heat. As such, it might be inferred that insects are significant casualties but we have no proof that this is the case.

What happens to soil-dwelling insect larvae during heatwaves and drought? Is there a loss of biomass and under what conditions does such a loss become significant? Remarkably, there is precious little research into the effects of such events on fly larvae, even though flies form a substantial proportion of the diet of breeding birds. There are plenty of hoverflies whose larvae develop in damp soil or in very shallow water bodies, but research is needed for a much wider range of taxa.

Surprisingly, although we know something about the levels of hoverfly activity during daylight hours, there are big gaps in the published literature for early morning and late evening (*i.e.* before about 8 am and after 5 pm). We know of one very good MSc thesis that investigated this issue but, as it was never published, the results are not widely available. This sort of project, combined with detailed investigation of hoverfly behaviour during hot (or cold) conditions, would throw light on aspects of climate influences that may help to explain the relationship between climate change and hoverfly abundance.

Inevitably, there is a strong case for use of modern DNA analysis to understand taxonomic conundrums better. For example, the *Dasysyrphus venustus* complex would benefit from further attention. There will doubtless be others. As we are primarily ecologists and not taxonomists, guidance on these aspects would be best coming from others!

DNA analysis may highlight anomalies in the known fauna. For example, we are aware of the potential addition of a further *Sphaerophoria* species on the basis of DNA analysis of a specimen for which the source of the DNA profile is unknown. Since the species concerned seems to be unlikely to occur in the location from which it was reported, this record had not been officially accepted at the time of writing.

Gardening for hoverflies

Gardens can be great places to observe hoverflies – in one Peterborough garden alone nearly 70 species have been recorded. This may in fact not be an unusual number, but since relatively few people monitor their garden intensively it is not possible to be sure.

The usual way of thinking about the use of the garden by invertebrates is to focus on nectar sources. This is as true for hoverflies as it is for butterflies, although the choice of plants differs. Hoverflies will visit butterfly lures such as *Buddleja* and *Hebe* but they also benefit from a broader suite of flowers throughout the season.

There are some simple things that can be done to make a garden more attractive to hoverflies. For example, since hoverflies love *Allium* flowers it is worth leaving leeks that have bolted to flower. Alternatively, consider planting some of the bigger ornamental *Allium* species to see what they attract. Parsnip flowers also provide an attractive nectar source, so try allowing a few less vigorous plants to bolt. Fennel, as well as most native umbellifers, such as Hogweed and Wild Angelica, are also potentially good lures. If you do not like the idea of encouraging them in your garden, try growing them in pots!

Ironically, one of the most attractive plants to hoverflies is one of the gardener's worst enemies – Ground Elder. Although it is invasive, if you have a big enough garden try not to eliminate entirely this fantastic nectar source. This illustrates an important point: many of the best hoverfly lures are wild flowers. As native shrubs are particularly attractive to hoverflies, consider planting a hedge of Blackthorn and/or Hawthorn, or growing dogwoods or Wild Privet (rather than its Japanese counterpart) as ornamental plants. Hoverflies will not confine their attention to native species and many ornamental plants can also provide good nectar sources. The list on *page 336* provides a few ideas and indicates the months in which they are usually in flower.

Many of the hoverflies that visit gardens have larvae that feed on aphids. It is therefore worth considering 'companion planting', an approach whereby flowers and vegetables are grown close together. The theory is that hoverflies attracted to the flowers' nectar sources lay their eggs on the nearby vegetables, where there are often aphid

Female Great Pied Hoverfly *Volucella pellucens* on a Blackberry flower (TOP) and Marmalade Hoverflies *Episyrphus balteatus* on a poppy (BOTTOM).

colonies. The resulting larvae feed on the aphids and provide a natural means of pest control.

Thriving populations of predacious species can also be encouraged by the provision of over-wintering sites for puparia. One good example is *Epistrophe eligans*, which is fond of aphids on plum trees. The larvae of this hoverfly need a dry, sheltered place on the ground to pupate, and benefit from some corrugated cardboard stuffed into a suitable waterproof receptacle that is open at one end (pointing downwards to avoid filling with rainwater).

Other garden features that are important for hoverflies could be maintained by avoiding modern silvicultural practices. For example, hoverflies that breed in decaying roots do not benefit from the action of the modern stump-grinder. Unfortunately, this tool is now widely used in many parks and gardens, resulting in the elimination of much prime hoverfly habitat.

Another way of encouraging hoverflies to inhabit your garden is to provide hoverfly nest boxes. An old bucket placed in a corner and filled with twigs and leaves, and topped up with water, will rapidly attract *Myathropa florea* whose larvae often occur in good numbers in such situations. For a more bespoke version, cut a hole in the side of a two-litre plastic drink bottle and tape it low down on the sheltered side of a tree (away from direct sunlight as this will cause water temperature to rise in the summer and risks

Mature larva of *Epistrophe*.

A 'nest box' for *Myathropa florea*. An empty plastic drink bottle with a hole cut in the side has been taped to a tree. The water in the bottom contains a handful of twigs and leaves.

Garden hoverfly monitoring
Even if you do not go far from your garden, you can make an important contribution to monitoring hoverflies as there are currently only two long-term datasets available. The studies that contributed to these data sets were undertaken by Alan Stubbs in his Peterborough garden (Stubbs, 1991) and by Dr Jennifer Owen in her garden in Leicestershire (Owen, 2010). The technique used by Alan Stubbs is based on the concept of the butterfly transect methodology devised and used by Butterfly Conservation. It involves recording the hoverflies seen along a standardised walk around the garden each week from April to October. Jennifer Owen ran a malaise trap in her garden from 1971 to 2000. Maintaining a daily log of species and numbers seen is more helpful than maintaining a record for individual weeks.

cooking the larvae). The bottle should be partly filled with sawdust, twigs and water and kept topped up during dry weather. Such artificial nests may also occasionally yield scarcer species than *M. florea*.

Garden plants that are attractive to hoverflies by flowering period
* indicates species, or groups that contain species that are native or widely naturalised

January–March
* **Winter aconites** *Eranthis* spp.
* **Crocuses** *Crocus* spp.
* **Oregon-grapes** *Mahonia* spp.
  (different species flower at different times from November through to March)
* **Laurustinus** *Viburnum tinus*
* **Willows** (sallows) *Salix* spp.

April–May
* **Cherry Laurel** *Prunus laurocerasus*
* **Cherry Plum** *Prunus cerasifera*
* **Mexican Oranges** *Choisya* spp.
* **Traveller's-joys/Clematises** *Clematis* spp, (early flowering species)
* **Dandelions** *Taraxacum* agg.
  (leave them in the garden)
* **Spurges** *Euphorbia* spp.
* **Forget-me-nots** *Myosotis* spp.
* **Fruit tree blossom**
* **Green Alkanet** *Pentaglottis sempervirens*
* **Lithodora** spp.
* **Pieris** spp.
* **Rosemary** *Rosmarinus officinalis*
* **Skimmia japonica**
  (male plants are best)
* **Tree Peony** *Paeonia suffruticosa*
  (single varieties)

May–June
* **Blackberry/Loganberry** and **Raspberry** *Rubus* spp.
* **Pot Marigolds** *Calendula* spp.
  (single varieties)
* **Camellia** spp. (single varieties)
* **Crane's-bills** *Geranium* spp.
  (herbaceous hardy Geraniums)
* **Dogwoods** *Cornus* spp.
* **Feverfew** *Tanacetum parthenium*
* **Marjoram** *Origanum majorana*
* **Pieris** spp.
* **Wild Privet** *Ligustrum vulgare*
  (not Japanese Privet!)

July–August
* **Yarrows** *Achillea* spp.
* **Anise Hyssop** *Agastache foeniculum*
* **Masterworts** *Astrantia* spp.
* **Bishop's Flower** *Ammi majus*
* **Shrubby Hare's-ear** *Bupleurum fruticosum*
* **Borage** *Borago officinalis*
* **Fennel** *Foeniculum vulgare*
* **Buddleja davidii**
* **Buddleja × weyeriana**
* **Canadian Goldenrod** *Solidago canadensis*
* **Oxeye Daisy** *Leucanthemum vulgare*
* **Sea-hollies** *Eryngium* spp.
* **Hebes** *Hebe* spp.
* **Japanese Anemone** *Anemone hupehensis*
* **Knapweeds** *Centaurea* spp.
* **Lavenders** *Lavandula* spp.
* **Leeks** *Allium* spp.
  (leave a few to run to flower)
* **Chinese Privet** *Ligusticum lucidum*
* **Marjoram** *Origanum majorana*
* **Mexican Sunflower** *Tithonia diversifolia*
* **Michaelmas Daisy** *Aster amellus*
* **Mints** (and relatives) *Mentha* spp.
* **Poppies** *Papaver* spp.
* **Scabiouses** *Scabiosa* spp.
* **Shasta Daisy** *Leucanthemum × superbum*
* **Argentine Vervain** *Verbena bonariensis*
* **Pearly Everlasting** *Anaphalis margaritacea*

September–October
White (or Heath) **Aster** *Symphyotrichum* [=*Aster*] *ericoides*
Symphyotrichum [=*Aster*] *lateriflorum* [=*lateriflorus*]
Tree-ivy × *Fatshedera lizei*
Japanese Aralia (or Japanese Fatsia) *Fatsia japonica*
Helenium spp.
* **Hemp Agrimony** *Eupatorium cannabinum*
* **Ivy** *Hedera helix*
Coneflowers *Rudbeckia* spp.
* **Butterfly Stonecrop** (or Ice Plant) *Hylotelephium* [=*Sedum*] *spectabile*

Further reading and useful addresses

Identification of the more challenging genera and species is beyond the scope of this book and requires the use of a full key to species with supporting descriptions and illustrations. For the British fauna, this is provided by Stubbs & Falk (2002), but other European works may also be useful: van Veen (2004) and Bot & Van de Meutter (2023) provide keys to the Western European species. These are somewhat more technical than Stubbs & Falk, but can be helpful if you get stuck! They also cover European species that could conceivably be found in Britain or Ireland. In the same vein, Martin Speight's project: Syrph the Net covers the entire European fauna and now includes many keys as well as a wealth of other information. This is published as a series of spreadsheets and document files that are distributed electronically.

For the identification of larvae, Rotheray (1993) is really the only source. Rotheray & Gilbert (2011) provides a great deal of information about the biology and ecology of the family and, as mentioned above, Syrph the Net is another useful source of information, including the wider, European distribution and status of species.

For more general information about finding and catching flies, dealing with specimens, *etc.*, *A Dipterists Handbook* (Chandler, 2010) is an invaluable source.

BOOKS AND PAPERS

BALL, S.G. & MORRIS, R.K.A. 2014. A review of the scarce and threatened flies of Great Britain. Pt 6: Syrphidae. *Species Status No. 9*: 1–124, Joint Nature Conservation Committee, Peterborough.
A detailed analysis of the trends within the British hoverfly fauna, combined with allocation of statuses following the latest IUCN guidance. Available as a PDF from www.jncc.gov.uk.

BALL, S.G. & MORRIS, R.K.A. 2020. Changes in the phenology of Britain's hoverflies (Syrphidae). *Dipterists Digest* (Second Series), **27**: 1–12.
Describes how the phenology of Britain's hoverflies has changed as winters have become warmer and shorter.

BALL, S.G. & MORRIS, R.K.A. 2021. Range expansion in British hoverflies (Diptera, Syrphidae). *Dipterists Digest* (Second Series), **28**: 59–87.
Describes three drivers of range expansion.

BALL, S.G. & MORRIS, R.K.A. 2022. A north-western shift in the range of the hoverfly *Leucozona glaucia* (Linnaeus) (Diptera, Syrphidae) in Great Britain. *Dipterists Digest* (Second Series), **29**: 151–167.
Describes the loss of *L. glaucia* from south-east England and possible range expansion in north-west Britain, with analysis of likely reasons.

BALL, S.G., MORRIS, R.K.A, ROTHERAY, G.E & WATT, K.R. 2011. *Atlas of the Hoverflies of Great Britain (Diptera, Syrphidae)*. Biological Records Centre, Wallingford.
This atlas is now largely out of date. Updated maps can be found on the Hoverfly Recording Scheme website www.hoverfly.uk/hrs.

BOT, S. & VAN DE MEUTTER, F. 2023. *Hoverflies of Britain and North-West Europe: A Photographic Guide*. Bloomsbury Naturalist, London.
A comprehensive and well-illustrated monograph on the hoverfly fauna of NW Europe, including Britain and Ireland.

CHANDLER, P. (ed.) 2010. *A Dipterists Handbook* (second edition). The Amateur Entomologist, Vol. **15**. Amateur Entomological Society, Orpington.
A comprehensive modern introduction to Diptera and their study.

CHANDLER, P. 2024. *An Update of the 1998 Checklist of Diptera of the British Isles*. British Entomological & Natural History Society.
Downloadable pdf available from www.dipterists.org.uk/checklist

GILBERT, F.S. & FALK, S.J. 1986. *Hoverflies*. Naturalists' Handbooks 5. Cambridge University Press.
Provides keys to selected species. This is an excellent guide to the study of hoverflies.

IUCN. 2012. *IUCN Red List Categories and Criteria*: Version 3.1. Second Edition. IUCN.

MORRIS, R.K.A. 2019. Understanding common misidentifications of British hoverflies (Diptera, Syrphidae). *British Journal of Entomology & Natural History*, **32**: 351–363.

MORRIS, R.K.A. & BALL, S.G. 2021. Death by one hundred droughts: Is climate change already driving biodiversity declines in Britain? *British Wildlife*, **33**: 13–20.
Discusses the implications of climate change for invertebrates.

ROTHERAY, G.E. 1993. Colour guide to hoverfly larvae in Britain and Europe. *Dipterists Digest* **9**. Derek Whiteley, Sheffield. (Out of print).
This is a must-have for the hoverfly enthusiast who wants to find and breed out hoverfly larvae.
www.diptera.info/downloads/df_1_9_Colour_Guide_to%20Hoverfly_Larvae.pdf.

ROTHERAY, G.E. & GILBERT, F. 2011. *The Natural History of Hoverflies*. Forrest Text, Cardigan.

SPEIGHT, M.C.D. 2011. Species accounts of European Syrphidae (Diptera), Glasgow 2011. Syrph the Net, database of European Syrphidae, Vol. **65**, 285 pp., Syrph the Net Publications, Dublin.

STUBBS, A.E. 1991. A method of monitoring garden hoverflies. *Dipterists Digest* (First Series) **10**: 26–39.
Still the only established technique for long-term surveillance monitoring of hoverflies.

STUBBS A.E. & FALK, S.J. 2002. *British Hoverflies: An Illustrated Identification Guide*. British Entomological & Natural History Society. (Revised and updated by Ball, S.G., Stubbs, A.E., McLean, I.F.G., Morris, R.K.A. & Falk, S.J.)
This is the essential companion to this Princeton **WILD***Guides* publication, as it includes keys to all British and Irish species, accompanied by illustrations of important features.

VUJIĆ, A. *ET AL*. 2022. *European Red List of Hoverflies*. European Commission.
Available as a PDF from
https://wikis.ec.europa.eu/display/EUPKH/European+Red+List+of+Hoverflies

VEEN, M.P. VAN. 2004. *Hoverflies of Northwest Europe*. KNNV Publishing, Netherlands.
Provides keys to many of the species that ultimately may occur in Britain or Ireland. This is an excellent book, but may not be suitable for the novice as it is not comprehensively illustrated.

JOURNALS

DIPTERISTS DIGEST
Published twice-yearly by Dipterists Forum. It frequently includes accounts of hoverfly ecology.
www.dipterists.org.uk/digest

HOVERFLY RECORDING SCHEME NEWSLETTER
Published as part of the *Bulletin of the Dipterists Forum*. Back numbers can be accessed from the Hoverfly Recording Scheme website.
www.dipterists.org.uk/hoverfly-scheme/home

USEFUL ADDRESSES

Centre for Environmental Data and Recording CEDaR
The official body to which records from Northern Ireland shoud be submitted.
https://bit.ly/3FjIh0i

Dipterists Forum
The national society devoted to the study of flies. The Forum has a membership of over 400 and is extremely active, running field meetings and training events: novices are encouraged to join experienced members to learn more.
www.dipterists.org.uk/home

Hoverfly Recording Scheme
Provides access to scheme outputs, news and announcements, and discussion forums.
www.hoverfly.uk/hrs

iRecord
One way of submitting records to the Hoverfly Recording Scheme. Records may only be accepted if they are accompanied by one or more photographs allowing identification to be verified.
www.irecord.org.uk

The National Biodiversity Data Centre
The official body to which records from the Republic of Ireland shoud be submitted.
www.biodiversityireland.ie

Steven Falk's Hoverfly Photographs
This is a large resource of reliably identified photographs of hoverflies with species accounts.
https://bit.ly/3YfcohV

SyrphBoard
An online platform for recording Hoverflies.
www.maploom.com/?page_id=1072

UK Hoverflies Facebook page
A very active group with members available to provide detailed advice on identification from photographs. (Available to Facebook users only.)
www.facebook.com/groups/609272232450940/

UK Hoverflies Larval Group
This is the 'go-to' group for those interested in hoverfly larvae.
www.facebook.com/groups/1580298322233838

Acknowledgements and photographic credits

This book is a greatly revised and expanded version of our original *Britain's Hoverflies*. We owe a great deal to the people who have made it happen. Above all, we would like to thank **Andy Swash**, who has not only had the challenging task of editing our text, but was also responsible for ensuring that the text fitted the allocated pages and the design layout. This was an extraordinarily demanding job and we know that Andy sweated blood and tears to deliver what we feel is a huge improvement on its predecessors. Equally, we are indebted to **Rob Still**, who developed the innovative initial design and is responsible for massaging the rough layouts that Andy created into the polished version that we see here; he has an uncanny knack of introducing the 'wow' factor into the design of field guides. **Steve Holmes** and **Martin Jones** also provided valuable feedback on the earlier editions as we sought to improve on what we had already produced.

As we have previously commented, had **Alan Stubbs** not encouraged us to take on the Hoverfly Recording Scheme in 1991, this book and so many other initiatives would not have happened. Alan's influence continues to this day, as he remains a huge source of wisdom and encouragement.

The book itself is the culmination of the efforts of a vast army of hoverfly enthusiasts, many of whom have only become involved in recording because the original WILD*Guide* acted as a stimulus to look at hoverflies. Today, we have a hugely active Facebook group and a greatly valued team of enthusiasts who help with species identification, data management and administration of the group. We are especially grateful to **Mick Chatman**, **Linda Fenwick**, **Adam Kelsey**, **Sue Kitt**, **Chris Sellen** and **Katie Stanney** in this respect.

The maps used in this version present a mixture of British and Irish data. Those data have been submitted by over 8,000 individual recorders, all of whom have played their part in making this book what it is. Until now, we have not been in a position to make any comment about Irish distribution but thanks to **Una FitzPatrick** at the National Biodiversity Data Centre in County Waterford, Ireland, and **Damian McFerran** at CEDaR in Northern Ireland it is now possible to offer a first depiction of Irish distribution.

Photographic credits

The production of this book would not have been possible without the help and co-operation of a large number of photographers. For each of the species that we have treated in detail, we did our best to find one or more field photographs that illustrate its identification features. These photographs are backed up by close-up shots of specimens to illustrate relevant features in detail. The specimens came mostly from the personal collection of Roger Morris. The photographs of specimens were taken by Stuart Ball and, for readers interested in macro photography, Canon equipment was used: an EOS 60D camera, MP65 macro lens and MT-24EX twin-flash unit. They are focus-stacked using a StackShot electronically controlled focusing rail and processed using Helicon Focus and Zerene Stacker software.

In the first two editions of this book, many of the field photographs were taken by **Steven Falk** and **Dr Brian Valentine**. Whilst we have continued to use Steven and Brian's photographs extensively in this edition, we have tried to broaden the range of contributors to reflect the vast army of photographic recorders who participate on the Facebook group. As a result, the total number of contributors has risen from 59 photographers to 105, and the total number of images featured has increased by more than 60%, from 650 to 1,048.

ACKNOWLEDGEMENTS AND PHOTOGRAPHIC CREDITS

The sources of larval photographs remain somewhat limited but in addition to Dr Brian Valentine and **Dr Graham Rotheray**, we must thank **Nicola Garnham**, **Dr Ellie Rotheray** and **Geoff Wilkinson** who have made it possible to broaden coverage of immature stages.

When we originally started searching for photographs online, we found many excellent pictures of hoverflies but a significant proportion were incorrectly identified. That situation has changed with the establishment of the Facebook group and we were able this time to conduct more effective searches using a database of URLs of Facebook posts. What we had not anticipated was that some photographers would post on Facebook, often for the purpose of confirming an identification, and then delete the original photograph! Sadly, this meant that we were not able to use a number of photographs that we would have liked to feature. The reliability of online identifications has improved markedly in the last ten years. Nevertheless, where others had made the initial identification, we made our own assessments of the photos we have chosen to use and take responsibility for the identifications given here.

The following is a complete list of all the contributing photographers, with details of their website where requested and the total number of images that are featured in the book. For readers wanting more information about any specific image, or to find out which images a particular photographer has taken, a comprehensive and fully searchable schedule of all the images is freely available for download from the Princeton University Press website at **https://press.princeton.edu/books/paperback/9780691246789/britains-hoverflies**. Particular mention must go to the following photographers who have ten or more photographs featured: **Steven Falk** [177], **Paul Kitchener** [72], **Paul D. Brock** [22], **Brian Valentine** [18], **Nigel Jones** [17], **Frank Vassen** [14] and **Ken Gartside** [10].

In total, four images are published under the terms of the Creative Commons Attribution 2.0 Generic license (CC BY 2.0) or the Creative Commons Attribution-ShareAlike 3.0 Unported license (CC BY-SA 3.0); these are indicated using the relevant code after the photographer's name in the list.

A **Ian Andrews** [1]; **Peter Andrews** [9];

B **Stuart Ball** [491]; **Tristan Bantock** [1]; **Vanna Bartlett** [1]; **Ian Beddison** [4]; **Colin Boyd** [1]; **Paul D. Brock** [22]; **Vic Brown** [1]; **Howard Burt** [1];

C **Graham Calow** [1]; **Mick Chatman** [1]; **Joan Childs** [3]; **Sean Clayton** [2]; **Richard Clifford** [1]; **Ashley Cox** [2];

D **Jelle Devalez** [1];

E **Jeremy Early** [1]; **Han Endt** [1];

F **Steven Falk** (www.flickr.com/photos/63075200@N07) [177]; **David Feige** [2]; **Linda Fenwick** [4]; **David Fotheringham** [3]; **Penny Frith** [1];

G **Nicola Garnham** [5]; **Ken Gartside** [10]; **Will George** [2]; **Madge Gibson** [1]; **Jennifer Gosling** [2]; **Janet Graham** (CC BY 2.0) (www.flickr.com/photos/149164524@N06) [3]; **Peter Greenwoods** [1];

H **Neil Halligan** [3]; **Andrew Halstead/Royal Horticultural Society** [1]; **Gail Hampshire** [1]; **Håkon Haraldseide** [2]; **Tim Harris** [1]; **Louise Hislop** [1]; **Martyn Hnatiuk** [3]; **Göran Holmström** [1];

J **Ron James** [9]; **Martin Jones** [4]; **Nigel Jones** [17]; **Steve and Gill Judd** [3]; **Maria Justamond** [1];

ACKNOWLEDGEMENTS AND PHOTOGRAPHIC CREDITS

K   Adam Kelsey [1]; Bob Kemp [3]; Roger Key [1]; Stephen King [2]; Philip Kirk [1]; **Paul Kitchener** [72]; Simon Knott [1]; Peter Krischkiw [1];

L   Jerry Lanfear [1]; Laurence Livermore [1]; Owen Llewellyn [1]; Richard Lyskowski [1];

M   Neil Marks [1]; **Tony Matthews** [9]; Harry McBride [1]; Ian McLean [1]; Roger Morris [5]; Mike Mullis [1];

N   Melanie Nichols [2];

O   John O'Sullivan [2]; Julian Oliver [2]; Jill Orme [5];

P   Alice Parfitt [1]; Mike Parnwell [1]; Martin Parr [1]; **Karl Parsonage** [8]; Adrian Plant [1]; Frank Porch [5];

Q   **Quartl** (CC BY-SA 3.0) (https://commons.wikimedia.org/wiki/File:Episyrphus_balteatus_qtl2.jpg) [1];

R   **Sandy Rae** [8]; Tim Ransom [2]; Linda Reinecke [3]; Jeremy Richardson [3]; **Ellie Rotheray** [6]; **Graham Rotheray** [6];

S   Chris Sellen [1]; Charles Sharp (www.sharpphotography.co.uk) [1]; Billy Smith [1]; Anne Sorbes [1]; Trevor Southward [5]; Paul Stevens [2]; Malcolm Storey [1]; Stephen Suttill [5]; Andy and Gill Swash [1];

T   Paul Tabor [5]; Pauline Taylor-Bradford [1];

U   Bill Urwin [1];

V   **Brian Valentine** [18]; **Frank Vassen** (www.flickr.com/photos/42244964@N03) [14]; Jaco Visser [4];

W   Mike Waite [7]; John Walters [4]; Walwyn [1]; Graham Watkeys [1]; Geoff Wilkinson [4]; Judy and Terry Wood [1]; Tim Worfolk [2]; and

Z   Cor Zonneveld [4].

The illustrations on *page 49* are by **Robert Still**.

A male *Rhingia campestris* extenting its mouthparts!

Index

This index includes the *scientific* names of all the hoverfly species mentioned.
Bold black text is used for species that are afforded a full account;
bold brown is used for tribes and sub-families.
Bold black numbers indicate the main species account.
Numbers in *italicised* text refer to the list of British and Irish hoverflies.
Numbers in regular text refer to other key pages.

ANASIMYIA 240, *308*
— *contracta* 240, *308*
— *interpuncta* 241, *308*
— *lineata* 240, *308*
— *lunulata* 241, *308*
— *transfuga* 241, *308*
ARCTOPHILA
(see SERICOMYIA) 266

BACCHA 74, *303*
— *elongata* 74, *303*
Bacchini 72, *303*
BLERA 278, *310*
— *fallax* 278, *310*, 322
BRACHYOPA 212, *307*
— *bicolor* 214, *307*
— *insensilis* 212, *307*
— *pilosa* 214, *307*
— *scutellaris* 212, *307*
BRACHYPALPOIDES
.... 280, *310*
— *lentus* 280, *310*
BRACHYPALPUS 282, *310*
— *laphriformis* 282, *310*

CALIPROBOLA 280, *310*
— *speciosa* 280, *310*
CALLICERA 164, *306*
— *aurata* 166, *306*
— *rufa* 43, 164, *306*
— *spinolae* 166, *306*, 322
Callicerini 164, *306*
CERIANA 108
CHALCOSYRPHUS 284, *310*
— *eunotus* 284, *310*
— *nemorum* 284, *310*
CHAMAESYRPHUS *252*
CHEILOSIA 170, *306*
— *ahenea* 174, *306*

— *albipila* 21, **184**, *306*
Cheilosia albitarsis **190**, *306*
— *antiqua* 174, *306*
— *barbata* 174, *306*
— **bergenstammi** **188**, *306*
— **caerulescens** **176**, *306*, 328
— *carbonaria* 175, *306*
— **chrysocoma** **186**, *306*
— *cynocephala* 175, *306*
— *fasciata* 192
— **fraterna** **186**, *306*
— *gigantea* 175
— *griseiventris* 174, *306*
— **grossa** 21, **184**, *306*
— **illustrata** **176**, *306*
— **impressa** **190**, *306*
— *lasiopa* 174, *306*
— *latifrons* 174, *306*
— **longula** **182**, *306*
— *mutabilis* 174, *306*
— *nebulosa* 175, *306*
— *nigripes* 174, *306*
— **pagana** **180**, *306*
— **proxima** **188**, *306*
— *psilophthalma* 175, *306*
— *pubera* 174, *306*
— **ranunculi** **190**, *306*
— *ruffipes* 180
— *sahlbergi* 174, *306*
— **scutellata** **182**, *306*
— *semifasciata* 22, 175, *307*
— **soror** **180**, *307*
— *urbana* 175, *307*
— *uviformis* 175, *307*
— **variabilis** **178**, *307*
— *velutina* 175, *307*
— *vernalis* 175, *307*
— *vicina* 174, *307*
— **vulpina** **178**, *307*

Cheilosiini 168, *306*
CHRYSOGASTER 206, *307*
— *cemiteriorum* 206, *307*
— *solstitialis* 206, *307*
— *virescens* 206, *307*
Chrysogastrini 196, *307*
CHRYSOTOXUM 102, *304*
— *arcuatum* 104, *304*
— *bicinctum* 102, *304*
— *cautum* 104, *304*
— *elegans* 106, *304*
— *festivum* 102, *304*
— *octomaculatum* 106, *304*, 322
— *vernale* 106, *304*
— *verralli* 106, *304*
CRIORHINA 292, *310*
— *asilica* 292, *310*
— *berberina* 294, *310*
— *floccosa* 294, *310*
— *ranunculi* 23, 292, *310*

DASYSYRPHUS 120, *304*
— *albostriatus* 120, *304*
— *friuliensis* 122, *304*
— *hilaris* 122, *304*
— *neovenustus* 122, *304*
— *pauxillus* 122, *304*
— *pinastri* 122, *304*
— *tricinctus* 120, *304*
— *venustus* 122, *304*
DIDEA 124, *304*
— *alneti* 124, *304*
— *fasciata* 124, *304*
— *intermedia* 124, *304*
DOROS 108, *304*
— *profuges* 108, *304*, 322
EPISTROPHE 136, *304*
— *diaphana* 138, *304*

Epistrophe eligans **136**, *304*
— *flava* **136**, *304*
— *grossulariae* **138**, *304*
— *melanostoma* **140**, *304*
— *nitidicollis* **140**, *304*
— *ochrostoma* **136**, *304*
Epistrophella euchroma 150
EPISYRPHUS **162**, *304*
— *balteatus* **162**, *304*
ERIOZONA **116**, *304*
— *syrphoides* **116**, *304*
Eristalini **218**, *308*
ERISTALINUS **234**, *308*
— *aeneus* **234**, *308*
— *sepulchralis* **234**, *308*
ERISTALIS **220**, *308*
— *abusiva* **232**, *308*
— *arbustorum* **232**, *308*
— *cryptarum* **228**, *308*, *322*
— *horticola* **230**, *308*
— *intricaria* **228**, *308*
— *nemorum* **226**, *308*
— *pertinax* **224**, *308*, *327*
— *rupium* **230**, *308*
— *similis* **226**, *308*
— *tenax* 25, **224**, *308*
EUMERUS **246**, *309*
— *funeralis* **246**, *309*
— *ornatus* **246**, *309*
— *sabulonum* **248**, *309*
— *sogdianus* **246**, *309*
— *strigatus* **246**, *309*
EUPEODES **146**, *304*
— *bucculatus* **146**, *304*
— *corollae* **148**, *304*
— *goeldlini* **146**, *304*
— *lapponicus* **150**, *304*
— *latifasciatus* **148**, *304*
— *lundbecki* **146**, *304*
— *luniger* **147**, *304*
— *nielseni* **146**, *304*
— *nitens* **146**, *304*

FERDINANDEA 23, **192**, *307*
— *cuprea* **192**, *307*
— *ruficornis* **192**, *307*

HAMMERSCHMIDTIA **216**, *307*, *322*
— *ferruginea* 25, **216**, *307*

HELOPHILUS **242**, *308*
— *affinis* **242**, *308*
— *groenlandicus* **242**, *308*
— *hybridus* **242**, *308*
— *pendulus* **242**, *308*
— *trivittatus* **244**, *308*
HERINGIA **260**, *309*
— *heringi* **260**, *309*
— *senilis* **261**, *309*

Lapposyrphus 150
LEJOGASTER **210**, *307*
— *metallina* **210**, *307*
— *tarsata* **210**, *307*
LEJOPS **244**, *308*
— *vittatus* **244**, *308*
LEUCOZONA **116**, *305*
— *glaucia* **118**, *305*, 331
— *laternaria* **118**, *305*
— *lucorum* **116**, *305*

MALLOTA **236**, *308*
— *cimbiciformis* **236**, *308*
MEGASYRPHUS **131**, *305*
— *erraticus* **131**, *305*
MELANGYNA **154**, *305*
— *arctica* **158**, *305*
— *barbifrons* **158**, *305*
— *cincta* **158**, *305*
— *compositarum* **156**, *305*
— *ericarum* **158**, *305*
— *labiatarum* **156**, *305*
— *lasiophthalma* **154**, *305*
— *quadrimaculata* **158**, *305*
— *umbellatarum* **156**, *305*
MELANOGASTER **204**, *307*
— *aerosa* **204**, *307*
— *hirtella* **204**, *307*
MELANOSTOMA **94**, *303*
— *certuum* **95**, *303*
— *dubium* **95**, *303*
— *mellarium* **95**, *303*
— *mellinum* **94**, *303*
— *scalare* **94**, *303*
MELIGRAMMA **150**, *305*
— *euchromum* **150**, *305*
— *guttatum* **152**, *305*
— *trianguliferum* **152**, *305*
MELISCAEVA **160**, *305*
— *auricollis* **160**, *305*

Meliscaeva cinctella **160**, *305*
MERODON **248**, *309*
— *equestris* 29, **248**, *309*
Merodontini **246**, *309*
MICRODON **298**, *310*
— *analis* **298**, *310*
— *devius* **300**, *310*
— *mutabilis* **300**, *310*
— *myrmicae* **300**, *310*
MYATHROPA **236**, *308*
— *florea* 24, 28, **236**, *308*
MYOLEPTA **217**, *307*
— *dubia* **217**, *307*
— *potens* **217**, *307*, *322*

NEOASCIA **200**, *307*
— *geniculata* **202**, *307*
— *interrupta* **202**, *307*
— *meticulosa* **200**, *307*
— *obliqua* **202**, *307*
— *podagrica* **202**, *307*
— *tenur* **200**, *307*
NEOCNEMODON **260**, *309*
— *brevidens* **261**, *309*
— *latitarsis* **261**, *309*
— *pubescens* **260**, *309*
— *verrucula* **261**, *309*
— *vitripennis* **260**, *309*

ORTHONEVRA **208**, *308*
— *brevicornis* **209**, *308*
— *geniculata* **209**, *308*
— *intermedia* **209**, *308*
— *nobilis* **208**, *308*

Paragini **96**, *303*
PARAGUS **96**, *303*
— *albifrons* **96**, *303*
— *constrictus* **96**, *303*
— *haemorrhous* **96**, *303*
— *quadrifasciatus* **96**, *303*
— *tibialis* **96**, *303*
PARASYRPHUS **142**, *305*
— *annulatus* **142**, *305*
— *lineola* **142**, *305*
— *malinellus* **142**, *305*
— *nigritarsis* 42, **144**, *305*
— *punctulatus* **142**, *305*
— *relictus* 144
— *vittiger* **144**, *305*

INDEX

PARHELOPHILUS 238, 308
— *consimilis* 238, 308
— *frutetorum* 238, 308
— *versicolor* 238, 308
PELECOCERA 252, 309
— *caledonicus* 253, 309
— *scaevoides* 252, 309
— *tricincta* 252, 309
Pelecocerini 252, 309
PIPIZA 256, 309
— *austriaca* 256, 309
— *bimaculata* 256
— *fasciata* 256, 309
— *fenestrata* 256
— *festiva* 256, 309
— *lugubris* 256, 309
— *luteitarsis* 258, 309
— *noctiluca* 258, 309
— *notata* 256, 309
PIPIZELLA 262, 309
— *maculipennis* 262, 309
— *viduata* 262, 309
— *virens* 262, 309
Pipizini 254, 309
PLATYCHEIRUS 75, 303
— *albimanus* 84, 303
— *ambiguus* 84, 303
— *amplus* 80, 303
— *angustatus* 90, 303
— *aurolateralis* 80, 303
— *clypeatus* 90, 303, 327
— *discimanus* 80, 303
— *europaeus* 80, 303
— *fulviventris* 92, 303, 330
— *granditarsus* 82, 303
— *immarginatus* 80, 303
— *manicatus* 86, 303
— *melanopsis* 80, 303
— *nielseni* 80, 303
— *occultus* 80, 303
— *peltatus* 88, 303
— *perpallidus* 80, 303
— *podagratus* 80, 303
— *ramsarensis* 80, 303
— *rosarum* 82, 303

Platycheirus scambus 80, 303
— *scutatus* 88, 303
— *splendidus* 80, 303
— *sticticus* 80, 303
— *tarsalis* 86, 303
POCOTA 296, 310
— *personata* 296, 310
PORTEVINIA 20, 192, 307
— *maculata* 192, 307
PSILOTA 250, 309
— *anthracina* 250, 309
PYROPHAENA 82

RHINGIA 194, 307
— *campestris* 194, 307
— *rostrata* 194, 307, 329
RIPONNENSIA 208, 308
— *splendens* 208, 308

SCAEVA 126, 305
— *albomaculata* 126, 305
— *dignota* 126, 305
— *mecogramma* 126, 305
— *pyrastri* 126, 305
— *selenitica* 126, 305
SERICOMYIA 266, 309
— *lappona* 268, 309
— *silentis* 268, 309
— *superbiens* 266, 309
Sericomyiini 266, 309
SPHAEROPHORIA 112, 305
— *bankowskae* 112, 305
— *batava* 112, 305
— *fatarum* 112, 305
— *interrupta* 112, 305
— *loewi* 112, 305
— *philanthus* 112, 305
— *potentillae* 112, 305
— *rueppellii* 114, 305
— *scripta* 114, 305
— *taeniata* 112, 305
— *virgata* 112, 305
SPHEGINA 198, 308
— *clunipes* 198, 308
— *elegans* 198, 308

Sphegina sibirica 200, 308
— *verecunda* 198, 308
SYRITTA 290, 310
— *pipiens* 290, 310
Syrphini 98, 304
SYRPHUS 132, 306
— *nitidifrons* 132, 306
— *rectus* 132, 306
— *ribesii* 132, 306
— *torvus* 134, 306
— *vitripennis* 134, 306

TRICHOPSOMYIA 264, 309
— *flavitarsis* 264, 309
— *lucida* 265, 309
TRIGLYPHUS 264, 309
— *primus* 264, 309
TROPIDIA 290, 310
— *scita* 290, 310

VOLUCELLA 270, 310
— *bombylans* 270, 310
— *inanis* 274, 310
— *inflata* 272, 310
— *pellucens* 272, 310
— *zonaria* 274, 310, 328
Volucellini 270, 310

XANTHANDRUS 92, 303
— *comtus* 92, 303
XANTHOGRAMMA
 .... 108, 306
— *citrofasciatum* 108, 306
— *pedissequum* 110, 306
— *stackelbergi* 110, 306
XYLOTA 286, 310
— *abiens* 286, 310
— *florum* 286, 310
— *jakutorum* 286, 310
— *segnis* 288, 310
— *sylvarum* 288, 310
— *tarda* 286, 310
— *xanthocnema* 286, 310
Xylotini 276, 310

About the authors

Stuart Ball and Roger Morris have together run the Hoverfly Recording Scheme since 1991. They are authors of several atlases of British hoverflies and the most recent *Status Review*. As active members of the Dipterists Forum, they have run many identification training courses across Britain. Both originally worked as entomologists for the statutory nature conservation agencies. Stuart Ball continued to work for the Joint Nature Conservation Committee until 2016. Roger Morris worked for English Nature/Natural England before leaving in 2009 to become an independent consultant. Both are now happily retired.